IMMANUEL KANT

IMMANUEL KANT

HIS LIFE AND THOUGHT

BY ARSENIJ GULYGA

TRANSLATED BY

Marijan Despalatović

BIRKHÄUSER

Boston • Basel • Stuttgart

Library of Congress Cataloging in Publication Data
Gulyga, Arsenij Valdimirovich.
Immanuel Kant.
Translation of: Kant.
Bibliography: p.
Kant Immanuel, 1724–1804. 2. Philosophers—Germany—Bibliography.
I. Title.
B2797.G8413 1985 193 [B] 85-7494

ISBN-13: 978-1-4684-0544-6 e-ISBN-13: 978-1-4684-0542-2
DOI: 10.1007/978-1-4684-0542-2

Softcover reprint of the hardcover 1st edition 1987

9 8 7 6 5 4 3 2 1

To Aleksej Fedorovič Losev

Table of Contents

Preface

Life and work of a philosopher cannot be unravelled into separate strands. The true events of such a life are only thoughts. Kant has no other biography than the history of his thought. He lived most of his life in one city — Königsberg, and never ventured beyond the frontiers of East Prussia. He did not seek fame, he did not gain power; neither in living, nor in loving, did he encounter turbulence. He never married.

Outwardly, Kant's life ran smoothly and uniformly. Perhaps it was more monotonous than the lives of other philosophers. That could not be said of the inner life, the life of the mind. There everything was wondrously alive. Daring ideas came into being, gathered strength, clashed with other ideas, yielded ground or ripened in the strife. Kant's thought roamed the continents, pushed beyond the confines of the earth, strove to the farthest reaches of the universe. His thought delved deeply into the human soul in order to know itself. It was intense and dramatic.

Almost all methods of modern philosophical enquiry are rooted in Kant. His ideas have undergone change, but they live on. A knowledge of Kant's philosophy is a good introduction to the study of philosophy in general. He teaches us to think independently.

Kant is compared to Socrates; for Kant's philosophy is profoundly human. The Athenian sage brought philosophy down from the heavens and gave it earth as its realm. He let the cosmos be and busied himself with man. For Kant, too, the problem of man is paramount. He does not forget the universe, but man is the fundamental issue. Kant's contemplation of the laws of being and consciousness has but one purpose: to make man more human, to make him lead a better life, to stop the bloodshed, and to prevent man's being confused by utopias and illusions. Kant calls all things by their proper names.

Kant was by no means the withdrawn recluse — a man not of this world. By nature he was sociable, by education and manners courtly. Very early there developed in him one life interest which extinguished everything else: philosophy. To this interest he yielded his whole being. For him, life was work, in work he found joy and meaning. Kant's life is an example of fusion of word and deed, of instruction and action. He died with a clear conscience, conscious of having fulfilled his duties.

The future philosopher was a sickly child; it was thought that he would lead a short and unproductive life. Yet he lived to a ripe old age, a rich life of work and health. He achieved that by the sheer power of his will. He developed a rigorous system of hygienic rules, adhered to them, and the results were brilliant. Kant made himself. In that, too, he is unique.

As long as his consciousness was clear and unhindered, Kant sought after the truth. But truth is a process. Kant never said that; he acted upon that assumption. He never deluded himself into thinking that the task had been accomplished, that the absolute had been discovered. He improved, refined, polished his thought. Kant's life is a ceaseless development in thought, an endless search — until the last years when he no longer had control of his thought.

It is difficult to read Kant. It is even more difficult to understand him. But once his thought has been grasped, it gives joy, it exhilarates. It is worth the labor — a mind strengthened by Kant's thought is rewarded a thousand times over. Take theses of a lesser author, break them down, free them of verbal subtleties: what is left is mere banality. In Kant, the difficulty of exposition always points to a difficulty in the problem as well, for Kant often posed totally new problems. Of simple things Kant speaks simply, but brilliantly.

The honour and the duty to write about Kant belongs to the philosopher. It belongs even more to a Russian, since there is a profound bond between Kant's thought and the hidden, fundamental thought of Russian classics. Two names should suffice — Dostoevsky and Tolstoy. They, as Kant, were moved by the fate of man; they, as Kant, saw the bottomless pit of possible conflicts, controversies, and catastrophies. Tolstoy — whom Hegel left cold — read and reread Kant. He was convinced that their views coincided. He collected Kant's aphorisms and published them. He said:

> "Kant's life never failed to move me deeply." He was asked, once: "Can an ordinary man grasp Kant's philosophy, and can it be made more accessible?" The answer: "Popularisation of Kant's thought would be a magnificent undertaking. I wonder whether they have done it in the West? It would certainly be highly desirable."[1]

This book does not claim to be an exhaustive and thorough study of Kant's philosophy; it is dedicated to his life. But Kant has no biography other than the history of his thought. Therefore it was not possible to avoid philosophy. The author strove to speak about major issues and — within the limits of the material — to speak plainly. If it could not always be done, the reader should be kind and forgiving.

— *ARSENIJ GULYGA*

Have the courage to use your own reason.
— KANT

CHAPTER ONE
Fruits of Enlightenment

We begin the biography of Kant in the traditional manner, with a history of the city of his birth. The granite of the city shaped, as it were, the rigorous architecture of his thought, the air of the city breathes in it . . .

Since about the eighth century there lived between the Vistula and the Niemen pagan Baltic tribes who called themselves Prussians. Christianity came to them with conquerors. After the failure of the Crusades in the Near East, the German knights turned eastwards: the Baltic was at their doorstep. Throughout most of the thirteenth century the German knightly orders laboured at subduing the Prussian lands.

In 1255 the cornerstone of a castle was laid at the mouth of the river Pregel. The knights named the castle Königsberg, in honour of the Bohemian king Ottokar II, their ally in the Crusade.

Gradually the land around the castle became settled. Three towns came into being: the oldest — Altstadt; to the east of Altstadt — Löbenicht, and to the south, on the island in the river, Kneiphof. The towns traded with one another, competed against one another, feuded among themselves. On occasion they made war on one another. Finally, in 1724, the towns consolidated into a single city. Kant and Königsberg were born in the same year.

The castle resembled a fortress more than a palace. It was built over centuries and in a variety of styles, a square edifice with a large inner courtyard, with many living quarters, stately ceremonial chambers and utility rooms; there was the large Moskowiter hall, one of the largest halls in the Germany of its day, and a tall watch-tower which first served a military purpose and then housed the fire-watch.

The inhabitants of the city came from all corners and ends of Germany. When Europe succumbed to religious wars, foreigners began to appear. Life in the war-torn land was hard and rude: one always had to be on one's guard and ready to do battle. Nature herself demanded strength and ceaseless exertion; marshes could not be turned into pastures, bitter winters ruined crops. Life taught the necessity of disciplined work.

Soon after 1517, the last Master of the Prussian Order left the Catholic church and introduced Protestantism in his country. In the beginning of the seventeenth century, the duke elector (Kurfürst) of Brandenburg became the Duke of Prussia; from then on, these two countries were governed jointly. In 1701, Duke Friedrich III proclaimed himself king (in, not of, Prussia). Berlin was the capital of the state, Königsberg was its largest city, the centre of eastern lands which lay isolated behind the possessions of the Polish crown.

The power of Königsberg grew quickly. The lively port was a bridge between eastern and western Europe. Crafts and trade flourished. A university (Albertina) was established and began to draw the youth of the Baltic lands who wished to study. There was also a strong garrison billeted not in barracks but in private lodgings, as was the custom of the day: the streets were always full of brightly colored uniforms.

A profusion of uniforms was characteristic of Prussian cities. In population Prussia ranked thirteenth in Europe, but in the number of soldiers it came fourth, right after France, Russia, and Austria. Military expenditures brought the young kingdom to the limit of its possibilities. In order to support a large army, the state had to levy heavy taxes and cut corners wherever it could. Valour was the first Prussian virtue, thrift the second. Princess Wilhelmina left a description of the frugal table of her father, king Friedrich Wilhelm I, from which guests got up hungry after conversing only about soldiers and thriftiness. The first interest of the "soldier-king" apparently overrode the second: parsimony did not deter him from spending large sums of money on the guards regiment of "lanky lads", giants recruited in all parts of the world. The tallest grenadier, an Irishman, cost the king 9,000 talers (a sum considerably in excess of the annual budget of the University of Königsberg). Everywhere else the king saved mightily, for instance on items of fashion. Expensive, clumsy wigs were banned, and men were ordered to tie their hair in simple queues. The fashion quickly spread throughout Europe. The king himself was always in uniform, another innovation. He saved on the sciences as well: he practically closed down the Academy which Leibniz had founded in Berlin under the king's predecessor.

Friedrich Wilhelm I left his mark in the history of German philosophy by ordering Christian Wolff out of Prussia within forty-eight hours and upon pain of death: the king had been informed that the freethinking doctrine denying the freedom of the will could be used to justify desertion. Of Leibniz, Friedrich Wilhelm spoke with scorn: in his opinion the fellow was not fit to stand guard duty. Once the king, in an expansive mood, laid his eyes upon the philosopher Edelmann and gave him a gulden; the thinker wanted to give two guldens to the king in return but bethought himself in time that the matter might end with caning. The only scientific experiment the "soldier-king" undertook was research in how to obtain strapping progeny from his lanky guardsmen. He married them to chosen maidens of great stature. The experiment, naturally, failed. The only academic debate held at the king's request was on the theme of whether all learned men were not windbags and fools.[2] The royal sense of humour was felt in the financial affairs of the Academy of Sciences as well. The Bible and the military service manual were the only books Friedrich Wilhelm read.

As often happens, the king's son was his antithesis in every respect. The crown prince was an enthusiastic musician, wrote poetry, loved French literature and philosophy. He found Prussia unbearable and tried to escape abroad but was caught and thrown into a fortress prison. His companion was beheaded before his eyes. The crown prince himself faced court martial and the firing squad for desertion. Everything, however, worked out in the end: in 1740 he ascended the throne of his father and as Friedrich II reigned for forty-six years.

The first seven years were the years of war with Austria. In the end, Prussia acquired Silesia, and her army added many batallions and squadrons to its ranks. No radical reforms were undertaken. "From the very beginning," Franz Mehring writes, "Friedrich understood that in harmony with the Prussian order of things each Prussian monarch must continue on the course set by his predecessors, that his place in history, his greatness — if one is to use the word — sprang from his steadfast refusal to go against the mainstream, even if his natural abilities and inclinations made the temptation to do it greater than that to which other Prussian kings had been exposed."[3] The flutist and poet who called his uniform his "shroud" became the symbol of Prussian militarism. Before accession to the throne he wrote "*Antimachiavelli*", a tract in which he characterised the monarch as "the servant of his subjects." Upon assumption of power, he embodied the principle of absolutism. By nature an Epicurean, he played the role of a stoic. The Francophile, enamoured of everything French, was to go to war with France.

But there were changes. On the throne sat a king well-read in philosophy, himself a writer of learned treatises, a freethinker who countenanced facetious remarks like: "May the almighty God, if there is one, have mercy on my sinful soul, if I have one." Voltaire was a guest at Friedrich's court; Lamettrie, hounded out of France and Holland, found refuge in Berlin where the king pampered him after he had appointed him to a position at court. "The protection Lamettrie enjoyed as the king's personal physician," Franz Mehring says, "and the beautiful obituary the king delivered in memory of the confirmed materialist in 1751, show that the level of the king's philosophical understanding was unequalled by any other German, including the young Lessing who was at that time fulminating against Lamettrie with more piety than wisdom."[4] Prussian despotism became "enlightened." Enlightenment gradually came into its own even there.

Enlightenment is an essential stage of cultural development of every country breaking away from feudalism. Enlightenment is fundamentally democratic: it wishes to bring culture to the people, the masses. It sees as its major tasks education and cultivation of sensibilities, together with dissemination of knowledge in the broadest possible sense. The Renaissance ideal of the free personality becomes in Enlightenment the attribute of universal validity. With it grows the sense of each individual's responsibility: the man of Enlightenment does not think only of himself but also of others and of his place in society. Social ideas take root: the problem of the best social order occupies the centre of interest and attention. The notion of equality sets minds on fire: not only equality before God (that is already stated in Christian doctrine) but equality before the law, universal equality. This equality is formal, but the civil legal system needs no other category. Enlightenment holds fast to the concept of formal right because in it it sees the guarantee of humanitarianism.

The Enlightenment sees knowledge made available to all as the remedy for all social ills and abuses. Knowledge is power: to acquire it, to make it universal is to put in man's hands the key to the secrets of his being. A turn of the key, the Sesame opens, and a rich life has been discovered. The possibility that knowledge might be misused is, on this agenda, out of the question. The early Enlightenment is rationalist, it is the age of understanding. Disappointment sets in quickly. Then man seeks salvation in "unmediated, direct knowledge," in feeling, in intuition, and somewhere ahead looms dialectical reason. But as long as the growth of knowledge remains a source of salvation, the ideals of Enlightenment remain unshaken.

The third characteristic of the Enlightenment is finally historical optimism. The idea of progress is an achievement of this period. Never before had an age thought of justifying itself to itself. Antiquity did not wish to know its predecessors, Christianity declared that its origin was decreed by the almighty. Even the Renaissance, which stepped forth as the arbiter of the dialogue between the two preceding cultures, saw as its task not a movement forward but a return to the source. The epoch of the Enlightenment was the first to develop a consciousness of itself as a new beginning. From that it was only a step to historical thought. And although not all proponents of the Enlightenment were able to achieve a historical view of things, the roots of historical thought lie in this period.

Proponents of the Enlightenment fought unremittingly against superstition, fanaticism, intolerance, deception, and general ignorance. They considered themselves missionaries of reason called forth to open man's eyes to his nature and his destiny; they were meant to improve man's lot and lead him on the path of truth.

It is not easy to determine the chronology of the Enlightenment. Feudalism gave way at different times in different countries. In Europe, Holland and England were in the vanguard. Then came France's turn. In Germany, the eighteenth century is the age of Enlightenment. Compared to England and France, Germany was backward: even there, however, capitalist relations of production were maturing in the womb of the feudal order. Agriculture held the price of place, but market economy was beginning to make itself felt. Capitalist manufacture grew in towns, trade flourished. Social changes were imminent everywhere.

The historical development of the German nation was at the time characterised by the economic and political fragmentation of the country. Broken up into many minature kingdoms, Germany did not make up a unified state. Formally there existed one German empire embracing all German lands (eastern Prussia was nominally outside of the empire), but actually each German monarch was the lord and master of his home. Austria claimed leadership in German affairs. The rise of Prussia brought a dangerous adversary into play. Progressive Germans who concerned themselves with the fate of their fatherland saw clearly that prosperity was possible only when feudalism had been eliminated and the country had been unified. The idea of national unity animated proponents of Enlightenment in all their endeavours, but in the eighteenth century the idea never turned towards the perversions of nationalism and chauvinism. All nations were equal as all men were equal: it was fashionable among the intellectuals to be a "citizen of the world," to whom the limitations of nationalism and its arrogant pride were foreign.

The philosophy of the German Enlightenment was formed not only under the sign of national movements but also under the influence of progress in the sciences. If Germany lagged behind England and France in matters economic and political, in the sciences that was not the case. German natural sciences were on the rise in the eighteenth century, they went through the same stages as were characteristic for European science in general. The accumulation of an enormous body of facts gathered and classified in the preceding period posed now problems of interpretation: it was necessary to consider nature in its living connexions and its development. The road to philosophy lay open.

The predominant school at first was that of Christian Wolff. When the soldier-king gave Wolff forty-eight hours to leave the country,

Wolff had held the chair at Halle for seventeen years, and a number of his works had earned him a European reputation. The philosopher knew the harsh nature of his monarch, so he did not wait for the period of grace to expire: twelve hours after the receipt of the royal order Wolff left town. The expulsion only enhanced the philosopher's fame: he received a chair at Marburg (where Lomonosov later studied under him). London and Paris elected him to their academies. Stockholm and Petersburg invited him to lecture; the Academy of Sciences was about to be established in Russia, and Peter I offered Wolff the post of its vice-president. The philosopher courteously declined. Nor was he in a hurry to return home, even after he had been rehabilitated by Friedrich Wilhelm I. Only the new king could prevail upon Wolff to return to Prussia. Friedrich II conferred upon him the title of privy councillor and gave him the then extraordinary salary of 2,000 talers. But Wolff chose not to accept the post in the capital and asked to be permitted to return to Halle.

Wolff's return to Halle was a triumphant occasion. Crowds met him well before the gates of the city, and an improvised procession ushered him in. At the head of the procession three mounted trumpeters blasted forth; then came fifty mounted students, the privy councillor Wolff with his wife in a coach and four, and behind them coaches of prominent citizens and university professors. When the procession reached the centre of the city, a band struck up and the music went on well into the night.

Wolff began to teach. But, as one semester followed another crowds of listeners began to thin out, and before long the famous teacher stood in front of empty auditorium seats. Wolff's followers took that to be the portent of victory: the master's teachings had penetrated so deeply into German education that is was no longer necessary to sit at his feet. Wolff's opponents held that he had simply outlived his fame: the fragile bubble had burst.

Compared to Leibniz, Wolff said little that was new and managed to say it crudely and poorly. Wolff's teleology, his doctrine of final causes, is generally characterised as "dull," even primitive. His contribution lay in the systematic exposition of Leibniz's philosophy; he gave Germany the first system embracing all fundamental areas of philosophical knowledge. In that respect he founded the first school of philosophical enquiry. His followers applied themselves with a will to the dissemination of scientific knowledge. Aiming at the broad reading public, the Wolffians developed what became known as "popular philosophy." They were convinced that education would gradually solve all the festering problems of the day. They worshiped reason, but maintained respect for the Christian faith, to which they sought to give a rational foundation. The centre of popular philosophy became Berlin. The term "Berlin Enlightenment" came into being.

Side by side with Wolffianism there existed in the German Enlightenment another movement, bound to the popular protest against social oppression and bitterly opposed to the official dogma of the church. The major source of these liberal ideas was Spinoza's philosophy. Spinoza's work had been making inroads in Germany since the end of the seventeenth century and overcoming the resistance of official protectors of ideological purity and of the moderate wing of the Enlightenment. Spinozism could not batter down the doors of the university. Spinozistic books were burned, their authors prosecuted. They were largely harmless, sometimes quite unknown men of letters who came, spread confusion, and departed without a trace. Books about freethinking were distributed in secret, sometimes achieving popularity, at other times remaining within the small circle of the initiated. Only after Lessing's death, for instance, did it become known that he had professed himself a Spinozist.

German freethinkers — unlike their French brethren — tended to be very careful in the matter of belief in God. The authority of religion remained firm. In the Middle Ages religion ruled supreme over the life of the spirit, it subjected to its power the sciences and the arts, it interfered in political affairs. Reformation had set Germany free of the rule of the Roman curia, although Luther and his followers were in no way inclined to lend support to freethinking. The Protestant orthodoxy developed a system of dogma and smoothly assumed the role played by Catholicism before it. The church was not prepared to relinquish control over the spiritual life of the country. The ruling classes gave the church their full support in the undertaking. The struggle for the Enlightenment and against orthodox Christianity was fought — by and large — under the banners of religious tolerance so that a better religion could be established.

A specifically German variant of revived Protestantism was Pietism. The movement came into being at the beginning of the seventeenth century as a protest against the spiritual stagnation and degeneration of Lutheranism. The beginnings of Pietism can be traced back to the great mystic Jakob Böhme. One of his followers, Quirinus Kuhlmann, who travelled far and wide to spread the teachings of his master, was burned at the stake as a heretic. Another—Jakob Spener—did rather better: he became the founder of the Protestant revival movement. Pietists rejected ritual and ceremony and placed the centre of religious life in the inner conviction, a thorough knowledge of the Bible, and personal piety and penitence. Pietism soon became intolerant: it turned into fanaticism and exalted asceticism. But in its time it was a breath of fresh air; many activists of the Enlightenment were nurtured on the spiritual values of Pietism and developed its radical, anti-dogmatic and anti-clericalist tendencies further. Prussia (Halle and Königsberg in particular) was the seedbed of Pietism.

At five o'clock on the morning of 22 April 1724, a son was born in the family of Johann Georg Kant, a harness-maker in Königsberg. According to the old Prussian calendar it was the day of St. Emmanuel; the boy was given the Biblical name which in translation means, "God is with us."

Kant claimed that he was of Scottish ancestry. As it turns out, the philosopher was wrong: his great-grandfather, Richard Kant, was of Baltic descent, born in Prökuls, in what is today Latvia.[5] Judging by the available documents, the great-grandfather spoke no German. Richard's son moved to Memel, mastered the craft of harness-making and passed the craft on to his son Johann Georg, who settled in Königsberg. Two of Richard Kant's daughters married Scotsmen, thus, perhaps, giving rise to the legend of Scottish ancestry. The future philosopher's mother, Anna Regina, was born in Nürnberg, the daughter of a saddler.

The boy grew up in the outskirts of town, amid craftsmen and shopkeepers, in an atmosphere of hard work, honesty and puritanical severity. He was the fourth child of the family. Anna Regina gave birth to nine children, of whom five survived. Immanuel Kant had an older sister, two younger sisters, and a brother, Johann Heinrich.

Immanuel was a fragile child. Anna Regina had by then lost two children and she tried very hard to nurture her son's physical and moral health, to develop his curiosity and imagination. "I will never forget my mother; she planted and cultivated the first seeds of goodness in me, she opened my heart to the impressions of nature; she awakened and broadened my ideas, and her teachings have had a lasting, beneficial influence on my life."[6] Thus Kant in the decline of his years.

The spirit of pietism ruled in Johann Georg's house. "People can say about pietism whatever they please. Enough! Those who took it seriously gave a spendid account of themselves. They mastered the highest human qualities, that serenity, joy, and inner peace that could not be disturbed by passion. They feared no need, they were indifferent to persecution, no quarrel could arouse them to anger or wickedness."[7] Kant remembered a legal battle between the saddlers' guild and the harness-makers' guild. Kant's father lost a great deal in the case, but he never permitted himself to utter an angry word against those who were responsible for his misfortune. It is difficult to tell whether that was indeed the case; it is important to note, however, that Kant thought that it was the case, that it remained in his memory as one of the first lessons in ethics the future great moralist learned. From his father Kant also learned to love work.

Upon the advice of the family pastor, Franz Albert Schultz, the eight year old Immanuel was sent to the state secondary school, the "Friedrichskollegium." Pastor Schultz was the principal of the school. Here the future philosopher spent eight years. He studied in the "Latin school," and his main subjects were Latin (up to twenty hours of Latin a week!) and theology (cramming of catechism). It is from this source that Kant's love of Roman poetry and his antipathy to the externals of religious cults grew. The parents wanted to see a pastor in the family but the youngster, spellbound by the talented Latin teacher Heydenreich, dreamt about dedicating himself to the study of classical languages.

Priestly ambitions could easily be destroyed by the monastic discipline of *Collegium Fridericianum.* This was a pietist school, severe and unrelenting: vacations were unknown. Classes began at seven o'clock in the morning, but pupils were expected to be at their places by six. Morning prayers took half an hour, and each class was preceded by a prayer. Classes ended at four o'clock in the afternoon. On Wednesdays and Saturdays there were elective classes in mathematics, music, French and Polish. Greek and Hebrew were required subjects (they were part of theological studies). Natural sciences and history were not taught. Immanuel's weak health hindered him in his studies, but quick-wittedness, good memory, and sheer diligence evened the odds. For a number of years he was at the top of his class. He graduated second from the top.

In the fall of 1740 the sixteen-year-old Immanuel Kant entered the university. But which of the schools? It is difficult to answer this simple question, because records do not indicate a student's chosen field of studies. The University of Königsberg had four schools (*Fakultäten*): three of them — theology, medicine and law — were considered the higher; philosophy was the lower. Pursuant to Friedrich Wilhelm I's directives, students could enroll only in one of the higher schools (the soldier-king needed civil servants, and we already know what he thought of philosophy). Kant's earliest biographers claim that he chose theology, following the wishes of his parents. There is a number of reasons why we should not accept that view. For instance, the subjects to which Kant devoted himself with great zeal. The pupil's fascination with philology gave way to a lively interest in physics and philosophy. It is quite possible that Kant enrolled in the school of medicine; he would eventually show a remarkable command of the field and even write a treatise on the diseases of the head.

Kant's new interests were awakened by a man who influenced his development even more than Franz Schultz and Heydenreich: Professor Martin Knutzen, who died at the age of thirty-seven. Had he lived, he would perhaps have become one of the great names in German philosophy. Knutzen is today known only as Kant's teacher, yet he became a professor at twenty-one. As a Pietist and Wolffian, Knutzen was greatly interested in the achievements of English natural sciences. Kant first heard the name Newton from Knutzen. Under his influence and with his help, Kant began to compose his own work on physics in the fourth year of his university education.

The work was slow. Kant lacked methodological skills and his preparation was inadequate. What made matters even more difficult was his financial situation. When he entered the university, Kant had left home. His mother was no longer alive (she had died relatively young, when Kant was barely thirteen), and his father was making ends meet with difficulty. Immanuel supported himself by tutoring. His well-to-do classmates helped him out, from them he borrowed shoes and clothes. He found strength in aphorisms: "I strive to subordinate things to myself, not to be in thrall of them," "Do not yield to misfortune, meet it head on."[8] Pastor Schultz helped him now and then. A maternal relative, a prosperous bootmaker, helped more often. There are documents to the effect that even grandfather

Richter helped defray costs of publication of Kant's first treatise, *Thoughts on the Correct Evaluation of Life Forces.*

Kant spent three years on the writing of this piece, and four further years on its preparation for publication. The date on the title page is 1746, but that was the date when printing began: the last sheets left the printery in 1749. The author sent copies to Albrecht Haller, the Swiss savant and poet, and to the mathematician Leonhard Euler, in Petersburg. These are the first known instances of Kant's use of the postal services. He received no answer.

Perhaps the subject matter of the work has something to do with the silence. Kant sought to mediate between the Cartesians and the Leibnizers on the issue of how to measure kinetic energy. According to Descartes, it was directly proportionate to speed; according to Leibniz, it was the square of speed. Kant stepped in and suggested that in some cases Descartes' formula was to be applied, in others, Leibniz's. In the meantime d'Alembert had worked out the solution to the problem six years before, in 1743, in the following formula: $F = mv^2/2$. Kant obviously knew nothing about it.

Things like that are not unknown in the history of science. Fifty years later Hegel claimed in his dissertation that there could be no unknown planet between Mars and Jupiter, although the planet Ceres had by then been discovered. Lessing celebrated Kant's mistake in an epigram:

> Kant took up a mighty cause:
> To tell the world what's what.
> He measured forces grandiose,
> But his own he did not.

Kant's youthful work is interesting not merely as a biographical datum but also as a contribution to science. For instance, contemporary cosmology might find Kant's notions about the relationship between the three-dimensional space and the law of universal gravity quite interesting. "It appears that three-dimensionality is the result of force with which substances of the world act upon one another. The force is inversely proportionate to the square of distance."[9] Kant does not develop the argument further — that he leaves to others.

The first example of Kant's writing is interesting stylistically as well. Le style — c'est l'homme. Kant's book is not written in Latin but in his mother tongue; the prose is good, remarkably clear and simple. The book breathes with youthful fire and self-confidence. Its motto, from Seneca, catches the eye: ***Nihil magis praestandum est, quam ne pecorum ritu sequamur antecedentium gregem, pergentes, non qua eundum est, sed qua itur*** (Seneca, De vita beata, Cap. I). Had he already chosen his way? Most definitely: "I have already chosen the road I intend to follow. I will keep to it, and nothing will entice me to leave it"[10] (Preface). In a work on physics, the declaration seems out of place. But it is significant: the young thinker cannot refrain from expressing everything which moves him. His writing will become more austere as his work progresses; but at first there is an exuberant flood of energy, thoughts, and words.

Kant's first work is a document of an epoch which sought to subject all accumulated prejudices to the judgment of reason. New times had brought in new authorities. If the authority of Newton and Leibniz stands in the way of truth, Kant said, then it must be set aside: reason is the only authority. No one is safe from error and the right to remark on this is held by all.

Just as David felled Goliath, so a neophyte thinker can on occasion and in one area of knowledge or another surpass a thinker who is far superior to him in the general breadth of knowledge. Kant had himself in mind: "The truth, which so many of the greatest masters of human knowledge had sought so hard, and in vain, presented itself for the first time to my reason."[11] Then, the youngster paused: had he gone too far? He liked the sentence, he let it stand, adding a qualification: "I will not venture to defend this notion, but I would not wish to renounce it, either." Something like a compromise had been struck.

This is a telling detail. In his first work Kant already showed not only that he sought truth at all cost, but also that whenever faced with mutually exclusive opposites he sought rational compromise. In this work he was trying to "unify" Leibniz and Descartes; in his mature years he would do the same with major philosophical movements. One of the central preoccupations of the future critical philosophy is already evident: it seeks to expose contradictions and yet permit them to co-exist, to overcome one-sidedness and to offer a fundamentally new solution while taking up all solutions in the experience which has now been expanded, not to vanquish, but to reconcile.

Kant spent almost seven years at the university. In 1747 he left the city of his birth, without taking the final examinations. But he did not wander far. He found employment as a tutor in three God-forsaken corners of East Prussia. First he went to the village of Judtschen, near Gumbinnen, where he taught the three sons of pastor Andersch. The original inhabitants of the area, Lithuanians, had been practically eradicated by the devastating plague of 1709; the population now consisted primarily of immigrants from the French cantons of Switzerland. The pastor, who was a Silesian, was compelled to adapt to his flock, who spoke a different language. Kant saw that "children of different nations" could live in peace under the same sky. Here he developed an interest in the culture of Lithuania which he never lost.

In the summer of 1750 Kant moved to the opposite end of the province, in the vicinity of Osterode. He became the tutor in the family of a lord of the manor. Again there were three youths in his charge, the sons of a major von Hülsen. The youngest, George Friedrich, had fond memories of his first teacher for many years. Was it perhaps Kant who put in the head of the future land-owner the idea that serfs should be set free? The young man would later do just that. It is obvious that Kant left seeds of a moral and judicious understanding of the human condition in the heart of his charge.

The third teaching post was in the family of Count Keyserling. Biographers cannot agree on whether Kant lived on the count's estate near Tilsit, or journeyed from Königsberg to the Keyserling castle, which was nearby. The first extant portrait of Kant is a drawing done by the countess Keyserling. The young beauty was passionately interested in philosophy; rumour had it that she was no less passionately interested in the visiting philosopher. It was alleged that the feeling was returned.

In the Prussian wilderness Kant mastered not only the art of teaching. Here was a sound school of everyday life, here he got used to being with people and came to know the ways of different classes. Books and an abundance of free time laid the foundations of the future scientific activity. Kant returned to Königsberg with a bulky manuscript on astronomy, originally entitled *Cosmogony, or an Attempt to Explain the Origin of the Universe, the Formation of Celestial Bodies and the Cause of their Movement in Accordance with the Universal Laws of Motion of Matter and Newton's Theory.*

In June 1754 Kant published a brief statement in two issues of a Königsberg weekly, announcing the forthcoming publication of the treatise. The essay had been written as an entry in the competition announced by the Berlin Academy of Sciences, on the theme, "Has the rotation of the earth caused changes to occur . . ." Ultimately Kant decided not to enter the competition, and the prize went to a candidate who gave a negative answer to the question posed. Contrary to the submission of the unworthy laureate, Kant came to the correct conclusion that the earth in its rotation is being slowed down by the tides of the oceans. Kant's calculations are wrong, but the idea is sound.

The problem is that the gravitational pull of the moon causes the seas to flow from east to west, that is, in the direction contrary to the rotation of the earth, and thereby exerts a braking effect. It is true, Kant said, that when one compares the slowness of the tides with the velocity of the earth, and the trifling mass of water with the gigantic proportions of the globe, it might appear that the influence of such a movement would have to be disregarded as trivial. But if one bears in mind that the process occurs ceaselessly and eternally, that the rotation of the earth is a free movement and that the smallest losses of velocity are permanent, it would be unthinkable for a philosopher to pronounce this slight effect insignificant.

At the end of summer of 1754 Kant published another essay, "The question of whether the Earth is Growing Old, considered from a Physical Point of View." Kant had no doubt that the earth, too, is subject to the aging process. Everything comes into being, achieves completion, and then slides into a decline. The earth is, naturally, no exception. Kant's treatment of the concrete geological processes is cautious: he is critical of hasty conclusions. Some of them remind him of the complaints of the old that the world is not what it used to be, that old virtues have run their course and yielded to new vices, falsehood and deception have replaced honesty; the old have vainly persuaded themselves that the heavens brought them forth in a time of supreme prosperity, they cannot imagine that after their death the world will be just as lovely as it was when they were born.

Both of these essays were preludes to the treatise on cosmogony. Its final title was: *A Universal Natural History and Theory of Heavens, or, An Attempt to Explain the Composition and Mechanical Origin of the Universe According to Principles of Newtonian Physics.* The treatise was published anonymously in the spring of 1755 and dedicated to Friedrich II.

The book was dogged by bad luck: the publisher went bankrupt, his shop was closed down by the courts, and the book came out too late for the spring fair. But that is not the reason, as some contend, for

Kant's not achieving a degree of fame in Europe as the creator of a new cosmogonic hypothesis. The book finally came out, its anonymous author was identified, and a favourable review appeared in one Hamburg periodical.

In 1761 J.H. Lambert repeated Kant's theses about the structure of the universe in his *Letters on Cosmology*. In 1769 the French astronomer Laplace formulated a cosmogonic hypothesis analogous to Kant's; neither Lambert nor Laplace knew of their precursor. Such was the tenor of the time: Kant was not familiar with d'Alembert's work in kinetic energy, in the West no one knew of Kant's contribution.

In the discussion of his cosmogonic system Kant was at pains to preserve the harmony between the system and the faith in the existence of God. The thinker held that there was no conflict between the claims of faith and his hypothesis. However, he also had to acknowledge that his views were very close to the views of classical materialists — Democritus and Epicurus. Kant, like Democritus and Epicurus, thought that the primeval state of nature was a chaos of dispersed elementary particles of matter, atoms. Epicurus suggested that there was a force which compelled atoms to sink; the proximity of this notion to Newton's theory of gravity is evident, and Kant based his theory on Newtonian principles. The pupil of the Pietists felt he had to defend his stand: "Even the most outlandish views which have gained currency among men, contain a grain of truth."[12]

In the seventeenth century, the natural scientists — among them Newton and Galileo — believed that heavenly bodies had been created by God. Although Kant distanced himself from the classical materialists, he actually applied the principles of scientific materialism (after Descartes) to cosmogony: "Give me matter, and I will build a world out of it! I mean, give me matter, and I will show you how a world must arise out of it."[13] Kant's formula sounds like an aphorism, but it contains the fundamental premise of the book: Kant actually demonstrated how purely mechanical forces could fashion our solar system out of the primeval chaos of elementary particles.

Although Kant dispensed with God as the master builder of the universe, he still saw in him the creator of that chaotic matter out of which, according to the laws of mechanics, the edifice of the world was fashioned. Another problem, the origin of organic nature, Kant did not try to solve scientifically. Can we say, he asked, give me matter and I will show you how to make a caterpillar out of it? The varieties of objects are so many and so complicated that we can trip up with our first step. The laws of mechanics are insufficient for an understanding of the nature of life. A sound thought; but having uttered the thought, the young Kant sought no other scientific solution to the question of the origin of organic life. It was only in his old age, when he was thinking about the workings of the brain, that he uncovered an explanation of the existence of a complex structure of reciprocal relationships in the organism.

The treatise on cosmogony shows the same emotionally charged manner which Kant had exercised in his work on "forces of life." The author is moved by the grandeur of the universe and tries to find an adequate expression for the poetic revelation; again and again he quotes from his favourite poets — Pope, Haller, Addison. His prose begins to edge towards poetry. For instance, Kant suggests that the reader imagine himself observing the sun from close by.

> Broad seas of fire open themselves to your gaze, with tongues of flame shooting into the sky; fierce storms whose fury whips up the rage of fire and forces it now to debauch beyond its shores and flood the plains above, now to return to its confines; scorched rocks that push their terrifying heads out of blazing gorges so that tides of the fiery element now cover them up, now leave them bare, thus creating sun's spots; dense mists that put out the flames, and others that blown upwards by winds form ominous clouds, pregnant with torrents of fire, come down as burning floods upon the blazing valleys from the heights above; the crash of elements; ashes of matter consumed by fire, nature in the embrace of destructive powers yet even in the terrifying moment of its disintegration bearing witness to the beauty of the world and purpose of creation.[14]

Le style — c'est l'homme, and the young Kant is a child of the Enlightenment. He is filled with its pathos. The wondrous sense of the harmony of nature and of his own powers of insight pours out in the flood of spirited words.

Stylistic felicities, however, do not overshadow the main concern — the problem of the origin of the universe. The treatise has three parts. The first is intended as an introduction. Here Kant considers the systematic construction of the universe. The Milky Way is not a disorderly, scattered swarm of stars; it is an embodiment of laws that rule the whole solar system. The galaxy is flattened, and the sun is near the centre of the plane. There are many such systems of stars in the universe, and the infinite universe is systematic in nature, all its parts bound together in a reciprocal relationship.

In Kant's day, six planets were known: Mercury, Venus, Earth, Mars, Jupiter, Saturn. The young thinker suggested that there might be other, yet unknown, planets behind Saturn. Uranus was discovered in his lifetime, Neptune in the nineteenth century, and Pluto in our own day.

The second part of the treatise deals with the problem of the formation of celestial bodies and galaxies. According to Kant, the following conditions have to be met in order for the universe to come into being: there must exist particles of matter of different specific density, and two forces — that of attraction and that of repulsion. The differences in densities cause the matter to bunch up, as it were, providing centres of attraction which pull lighter particles towards themselves. As particles fall upon the central mass, they heat it up until it becomes viscous. Thus the sun came into being, as Kant described it so beautifully.

The force of repulsion, forever negating the force of attraction, prevents all particles from gathering in one place. The tension between the two opposing forces imparts to some particles rotation, thus creating new centres of attraction — the seeds of planets. Their satellites come into being in an analogous fashion. The same forces and the same principles are at work in other stellar systems. "All the fixed stars the eye can see in the bottomless depths of the heavens, seemingly scattered without rhyme and reason, are suns and centres of their own systems. By analogy, there is no doubt that these systems came into being just as our own did, from the smallest particles of elementary matter which filled the empty space, the boundless abode of God's being."[15]

The creation of the universe is not a matter of a moment, but of an eternity. It had a beginning, but it has no end. It may have taken the nature which envelops us millions of years or centuries to achieve the present degree of completeness. Millions, and "whole mountains of millions of centuries" will pass, cradling new, and ever more complete worlds. And the old worlds will die, just as countless living organisms die before our eyes every day. Nature is rich and prodigal: its capacity to create the simplest, as well as the most complicated systems, and destroy them, is inexhaustible.

Kant's universe is constantly expanding. Heavenly bodies closest to its centre develop more quickly but are also closer to extinction. New worlds are coming into being in the meantime at the periphery of the universe. Kant foretells the death of our planetary system. The sun, growing ever hotter, will in the end consume the earth and its other satellites by fire; it will break them down again to constituent elements and scatter those throughout the universe so that they may again take part in the creation of new worlds. "Through the endlessness of time and space we follow this phoenix of nature, who seeks death in its own flames in order to arise renewed out of its ashes."[16]

The third part of the book contains "An Attempt at a Comparison of Inhabitants of Different Planets." Educated men of the eighteenth century were quite certain that heavenly bodies were inhabited. (Newton thought that the sun itself was inhabited.) Kant is convinced that intelligent life exists in the universe, but he has a single reservation — not everywhere. Just as there are desert areas of the earth incapable of sustaining life, so there are uninhabitable planets in the universe. To those who hold that mankind is unique, the philosopher offers a satirical story: those creatures who inhabit the forest atop a beggar's head had long ago come to regard their residence as a gigantic world, and themselves as the crown of creation, until one of them spied the head of a nobleman. It immediately called all its neighbours together and said with much pleasure: we are not the only forms of life in nature; look, there is another land, even more densely populated.

Kant wishes to judge without prejudice. The endlessness of creation embraces everything in nature which it has brought forth out of its inexhaustible treasury of riches. Everything from the highest form of thinking life to the lowest insect is indispensable to the beauty of the whole. And man, who is still so far from the higher degree of perfection, must not have any illusions about that. Has one of the denizens of the beggar's hair ever wrought such slaughter among his fellows as Alexander the Great?

And so, Kant argues, the majority of the planets are inhabited, and those that are not will be in time.

> Should the immortal soul of man, whom the grave itself cannot annihilate but only change, remain forever chained to this speck in the cosmic space, this our earth? Is it never to get a closer look at the remaining wonders of the creation? Who knows whether

it is not ordained that it will come to know intimately the perfection of those distant cosmic bodies which so mightily arouse its curiosity? That is, perhaps, why new planets are coming into being, so that when the time given us to spend here has run out we may find abode under different skies? Who knows whether the purpose of Jupiter's satellites is not to become our future suns?[17]

What is the meaning of this? Is this a prophecy of space travel, or an attempt to bring the enquiring mind of a naturalist into harmony with the ruling dogmas of the church? The latter, more likely. Kant's scientific work rested securely on the foundations of a Pietist education. Fusion of daring scientific guesses with semi-mystical visions was very much the mark of the time. I beg the reader not to forget the quoted passage: three decades later Kant will read a similar statement in Herder, become indignant and ridicule his pupil.

Next Kant tackles the problem of how the distance from the sun influences the ability of living beings to think. The inhabitants of Earth and Venus cannot exchange their environments and survive: the earthling is made of matter adapted to a certain temperature; heat would shrivel and evaporate his organism. The Venusian would freeze into immobility in the colder reaches of the heavens. The Venusian's body must be made of lighter, more viscous matter than the body of the earthling. Only if that is the case can the weaker influence of the sun confer upon him the degree of mobility found in organisms on other planets. And then Kant formulates a law: the further a planet is from the sun, the lighter and finer the organic composition of its inhabitants.

The power of the soul depends upon the composition of the matter. If only thick juices course through the body and its tissues are coarse, its spiritual powers are reduced. And immediately we have yet another law: the further from the sun a planet is, the more beautiful and perfect are its inhabitants. Human nature, occupying the middle range of the living spectrum, sees itself between the two extremes of perfection. If the sight of rational creatures on Jupiter and Saturn fills us with envy, the sight of lower orders of development on Venus and Mercury will restore our spiritual equilibrium. What a wonderful spectacle — exclaims the philosopher. On the one hand thinking beings to whom a Greenlander or a Hottentot would appear a Newton, and on the other hand beings who would look at Newton with wonder, as we look upon an ape.

When one reads these lines, Voltaire's *Mikromegas* comes instinctively to mind. The hero of Voltaire's philosophical pamphlet is an inhabitant of the planetary system of Syrius, twenty-four thousand times taller than man, who travels through the universe accompanied by a dwarf from Saturn, who is only one thousand times taller than man. Eventually they end up on the earth, which to them appears a wretched place. After much exertion and with the help of a microscope the visitors discover on our planet rational creatures and enter into conversation with them (they are members of a French scientific expedition under the astronomer Maupertuis, measuring the earth's meridian). They quickly become convinced of the natives' staggering stupidity which, however, does not prevent them from carrying out exact scientific calculations.

If *Mikromegas* had not appeared three years before Kant's *A Universal Natural History and Theory of the Heavens*, one might think that Voltaire was parodying Kant. The "forces of attraction and repulsion" are mentioned in the pamphlet, as is the Englishman Derham, whose work, according to Kant, first gave him the idea of considering the systematic nature of the universe. All these coincidences are more than mere coincidences. Voltaire spent the early 1750s in Prussia, the recipient of countless signs of royal favour; *Mikromegas* made fun not only of the scholastic strife among the different schools of philosophy, but also of specific tendencies of the "Berlin Enlightenment." Everybody was mad about astronomy. Maupertuis, brought to Prussia from Paris, stood at the head of the Prussian Academy of Sciences. Kant did not live in Berlin, but he did live in the intellectual atmosphere of the capital, and his treatise reflected its strong and its weak sides. That is why today *Mikromegas* sounds like a satirical paraphrase of Kant's pages. Besides, Voltaire is better known than many who have been forgotten, but who in their day held sway and commanded respect and whose work was far more fantastically speculative than Kant's enterprise. For instance, Lambert claims in his *Cosmological Letters* that the most highly developed rational creatures live on comets.

A good deal of *A Universal History* today appears dated, if not outright funny. Contemporary science no longer accepts Kant's fundamental hypothesis, namely that the solar system was formed out of cold particles of elementary matter, nor any other of Kant's hypotheses for which he sought to offer adequate substantiation. But the central philosophical idea — the concept of a universal historicity, the idea of evolution — remains unshaken.

The man who formulated a theory about the origin of the universe and opened the way to a dialectical view of the world, the author of two books and two original treatises which had attracted attention, was still, technically, a student, or, more precisely, a candidate about to sit for his MA (*magister*) examinations. When he thought about his future, Kant saw himself as a teacher in a university. In the course of his tutorial employments he had set aside a certain sum of money which was absolutely essential for the launching of a university career. Everything was ready, all he needed was the degree.

Shortly before his thirty-first birthday, on 17 April 1755 Kant submitted his dissertation, *On Fire*, to the philosophical faculty. It is a manuscript of twelve calligraphically written pages, in Latin. The dissertation was not defended, it was significant only as the admission to the examination. The dissertation was accepted, and four weeks later Kant took his oral examination. Finally, on 12 June, there took place a festive promotion of the candidate to the academic rank. The dean delivered a discourse on a problem in Judaic studies, then the new *magister* gave a lecture, in Latin, and at the conclusion he gave thanks to the learned gathering, the men who had opened for him the doors to wisdom.

But, he was not yet a member of the faculty. In order to obtain *venia docendi*, the *magister*, or doctor of philosophy (the titles were identical), had to pass habilitation, i.e., the candidate had to submit and defend another dissertation. Furthermore, the candidate had to defend his work in public. Therefore, the title page of Kant's second dissertation runs as follows:

A Re-consideration
of the First Principles of Metaphysical Knowledge
Offered by Magister Immanuel Kant of Königsberg
Who Will
With the Permission of the Honourable Members of the
Philosophical Faculty

Defend the Said Re-consideration in a Public Debate
on 27 September 1755 from 8 until 12 o'clock in the Morning
in the Auditorium of the Philosophical Faculty
in Order that He May Become Member of the Said Faculty.
Respondent:
Christoph Abraham Borchard of Heiligenbeil, Prussia
Student of Theology
Opponents:
Johann Gottfried Moller of Königsberg
Student of Theology;
Friedrich Heinrich Samuel Lysius of Königsberg,
Candidate of Laws,
and
Johann Reinhold Grube of Königsberg,
Candidate of Laws.

The respondent in a debate supported and strengthened the arguments of the candidate. His role in the defense was important; Kant's dissertation was dedicated, in Borchard's name, to the governor of East Prussia, Lehwaldt.

The debate took place at the appointed time, and Kant obtained the post of adjunct assistant professor (*Privatdozent*), a member of the faculty whose salary was not paid by the state but by his students. Soon after his habilitation, Kant delivered his first lecture. The auditorium in the Albertina was not large enough, so many professors taught at home. Kant was a lodger in the house of Professor Kypke, in whose house one room was set up as a lecture hall. There the new *Privatdozent* gave his first lecture. There were so many listeners that not all of them could be accommodated in the room; some stood in the corridor and on the stairs. Kant was very nervous, he mumbled through the first hour and only after the break managed to get a grip on himself. Thus he embarked upon a forty-one year career as a university lecturer.

Half a year later, Kant applies for an adjunct professorship, that is to say a post without remuneration. There was nothing unusual in that: his teacher Knutzen had obtained this post when he was twenty-one, and since his death the post had remained vacant. In April of 1756 Kant petitioned Friedrich II to be given the vacant post. To this end Kant wrote a third required dissertation in Latin (*Monadologia physica*). He defended the dissertation successfully (one of the opponents was the sixteen-year-old Borowski, his future biographer), but the effort was in vain: the state decided to abolish the vacant post. Kant would wait fourteen years to become a full professor.

In his first autumn semester Kant taught logic, metaphysics, natural sciences, and mathematics. Later he would add physical geography, ethics, and mechanics. While a *magister* he had to teach four to six subjects a semester. Each week he taught at least sixteen hours, at most twenty-eight (that included a few hours of laboratory exercises). Here, for instance, is the agenda for one of his busiest days: from eight o'clock until nine — logic; from nine until ten — mechanics; from ten until eleven — theoretical physics; from eleven until twelve — metaphysics; after the midday meal, from two o'clock until three — physical geography; from three until four — mathematics.

"Every day I sit behind the anvil of my lectern and hammer my way through my lectures that are very much alike,"[18] Kant moans. Small wonder that in the second half of the 1750s he writes hardly anything: lectures exhaust his time and his strength. But, he no longer has financial worries. The *Privatdozent* can afford a servant: he takes a retired soldier, Martin Lampe.

Kant was especially proud of his course in physical geography. Geography, he said, is the foundation of history. Kant was one of the proponents of geography as an independent discipline. In the teaching of other subjects he used available textbooks (although as time went on he interpreted them more and more in his own fashion). There were no textbooks for geography, nor any teaching aids. Nor could Kant supplement the teaching with memories from his own travels. The gap was filled with readings.

He could offer lively and exact descriptions of foreign lands because he was possessed of a good memory, a lively imagination, and an appreciation of details coupled with an ability to shape details into wholes. Although he never left his study Kant undertook journeys around the world, sailed the seas and oceans, made his way across deserts. The teacher of geography had never seen a mountain range, yet his accounts were as forceful as if he had climbed them.

"I scoured sources, inspected everything I could lay my hands on . . . I read the best and most solid descriptions of distant lands."[19] But — the sources were meager, the accounts unreliable, the descriptions incomplete.

Here, for instance, is everything Kant could tell his listeners about Russia (as far as we can judge from the published text):

> The Asiatic parts of this realm are quite different geographically from its European portion. Gmelin believes that it is possible to consider the river Yenisey the physical border, because east of it the lay of the land changes drastically, the country becomes hilly, flora and fauna are unlike anything one finds to the west of the river. For example, here one finds musk. The sturgeon, a Volga river fish, swallows stones each year when the waters are in flood, and the stones, acting as ballast, enable it to remain close to the bottom of the river. The only difference between the sterlet and the sturgeon is in the taste of their meat: the meat of the sterlet is more delicate. In the Troitse — Sergheievsky monastery and near Kiev there are naturally preserved corpses of what are falsely considered martyrs.[20]"

Kant's information about Siberia is equally strange. Nowhere else in the world, he avers, is enjoyment of alcohol so widespread (the Muslim population is an exception, because their religion forbids the consumption of alcohol). The snow in winter is so deep that men fasten boards onto their feet in order to be able to move; they not only smoke tobacco, they also eat it. Grusia is, according to Kant, "a nursery of beautiful women." In Migrelia, he claims, it rains without end; the soil is so soft that it does not have to be ploughed before sowing.

And this is still a harmless anthology. True, Kant is sceptical of Pliny's accounts of one-eyed and one-legged peoples. But reports of men "with small, monkey-like tails" who live in the primeval forests of Formosa, Borneo, and the Orenburg steppe he considered quite probable. The publisher of Kant's lectures on geography (the publication took place towards the end of the author's life and he made no corrections in the text) found himself compelled to express reservations about the authenticity of such statements: they were indicative, he writes, of the level of knowledge in the eighteenth century.

Mistakes, however, cannot diminish the significance of Kant's work: for his time he gave an impressive general description of the face of the earth. He touched upon everything, the flora, the fauna, the mineral realm and the life of the peoples of four continents — Asia, Africa, Europe, and America. His discovered the mechanism of the formation of winds — the trade-winds and the monsoons. To anticipate a little, we can say that Kant's work in geography was largely responsible for his election to the membership of the Petersburg Academy of Science.

Kant is still fascinated by the natural sciences, but now there is also something new — an interest in philosophy. Kant's first purely philosophical work was his dissertation, "A Re-Consideration of the First Principles of Metaphysical Knowledge." In it Kant pursues Leibniz's principle of sufficient reason (*principium rationis sufficientis*). He draws the distinction between the reasons for the existence of an object and the reasons for knowledge of that object, that is to say, the distinction between the real and the logical reasons. The real reason for the set speed of light, for instance, is the composition of ether. Observations of Jupiter's satellites made the knowledge of this phenomenon possible. It was noticed that whenever Jupiter was farthest from the earth, the eclipse of its satellites deviated from the mathematically predicted norm — it was late. From that followed that light travelled through time, and it was possible to calculate the speed of light. The seed of a future dualism lay in such considerations: clearly, the world of appearances and the world of cognition are not identical.

Kant brought the principle of sufficient reason to bear on human action. And so there rose before him another problem which would concern him for the rest of his life — the problem of freedom. Even at this point the author was not indifferent to it, as it is clear from the manner of exposition: into the Latin text, composed according to rigid rules, there suddenly burst a spontaneous dialogue. Titus, the Wolffian, argued with Gajus, a follower of Crusius.

We remember how Wolffianism was explained to the soldier-king: since will is not free, a deserter cannot be brought before the court-martial because he is not responsible for his deed. Crusius' argument *contra* Wolff, which Kant then set forth, concludes as follows: if everything has a determining cause then we cannot be held responsible for our lapses, for God is the sole cause of everything. We therefore merely enact what has been preordained. Freedom of the will and determinism are mutually exclusive.

Kant is of the opinion that the notion of the determining cause is not in contradiction to the notion of freedom. Crusian free will designates only chance solutions, it lacks firmly established motives. It is the freedom of a game of dice, depending upon whether an even or an odd number comes up; the freedom of deciding whether to step out first with the left or the right leg. Kant's understanding of freedom is different — it is a conscious determination of an action in which reason relays the motive to the act of will. The problem of responsibility, of obligation, appears then as a question of "accountability," that is to say of the level, the clarity of consciouness. Evil exists in the world that God has created, but responsibility for evil done is exclusively man's. The Wolffian Titus says in Kantian dialogue that to be free to act is to act in accord with one's impulses, and that means consciously. Kant will later conclude that man must not act in accord with his impulses because they may suggest a course of action which is purely pleasing. All impulses are determined by nature; to act in consonance with them is to behave like an animal.

In general Kant has the same viewpoint as Leibniz and Wolff, although in several important details he departs from their point of view. Kant seeks a compromise again, this time between the metaphysics of Leibniz and Wolff, and the physics of Newton. He is not happy with Leibniz's notion that there exists between the body and the soul a pre-established harmony, that these two substances act in a pre-determined, synchronous, and yet independent manner. Newton's notion of reciprocity is much more satisfying to Kant. That being is harmonious by nature and that it irresistibly strives towards the universal good Kant does not as yet doubt.

Others, however, are beginning to have second thoughts. In 1753 the Berlin Academy of Sciences announced a competition for the best treatment of Alexander Pope's optimism. The contemporaries saw in the invitation an attempt to undermine Leibniz and Wolff. Under the protection of the Francophile king the Academy was packed with Frenchmen, who had brought in the spirit of scepticism, but no one dared attack Leibniz, the founder of the Academy. That is why the English poet, who in his famous *Essay on Man* had given a lyric expression to the Leibnizian conception of the best of all possible worlds, seemed like the perfect target of a possible critique. The laureate of the competition was a Crusianer, Reinhardt, who sought to prove that there may exist another world, no less perfect than the present one.

It did not end with the official award ceremonies. A polemic started in the press. The nature took a hand in the dispute. Towards the end of 1755 there occurred a tragic event which shattered the spiritual life of Europe, long since enveloped in repose, peace, and well-being.

An earthquake of frightening intensity fell upon Lisbon. There had been earthquakes before, but this time the catastrophe levelled a city in bloom, the capital of a European state.

Eyewitnesses gave horrifying accounts of the event. The sea, it seemed, suddenly boiled up; a gigantic wave came up over the port, tearing ships from their anchorages and tossing them onto the shore. The royal palace collapsed and sank in an instant into the sea. Churches fell apart like houses of cards. The earth spewed fire, within a few minutes tens of thousands died, hundreds of thousands were left wounded and homeless.

Many years later, Goethe drew a picture in *Poetry and Truth* of the spiritual and intellectual turmoil of those days.

> Never before had the demon of terror so quickly and overwhelmingly spread its horror over the earth. The boy, who was to hear of this time and again, remained shattered. God who made the earth and the heavens, who according to the first teachings of religion is so wise and merciful, had proved himself less than a loving father, dealing death to the just and the unjust impartially. In vain did the young soul try to establish the equilibrium swept away by these impressions. There was no help from the learned and wise writers who could not agree about how one should regard such a phenomenon.[21]

Voltaire reacted poetically to the destruction of Lisbon — a tragedy, unparalled since those of Herculaneum and Pompei. His verses rang out against optimism. Does God need such misfortunes? Are they punishment for sins? And is Lisbon more depraved than London or Paris? What were the sins of infants cleaving unto their mothers' breasts? Voltaire could find no answer, so he let his sarcasm loose upon the concept of fatal optimism. He would ultimately deal with it in *Candide*. In their passage through an unrelieved series of calamities the heroes of the story suffer through the catastrophe of Lisbon as well; although Pangloss, an adherent of the theory of pre-established harmony, remains oblivious to the world around, the reader can put two and two together. Kant, like his contemporaries, read and liked *Candide*.

But Kant was of a Panglossian persuasion himself, and although the Lisbon earthquake forced itself upon his mind he did not change it. On the occasion of the natural catastrophe he published two articles and a slim volume. The interest in the event was so large that the booklet was sold out literally page by page, as sheets came out of the printer's. Kent contemplates the human tragedy with the cool gaze of a realist. He begins the treatment of the earthquake on a cautionary note:

> By the history of the earthquake I do not mean the history of men's suffering, nor shall I give a list of devastated cities or their inhabitants buried under the rubble. One must draw upon one's imagination to represent however inadequately the terror that struck men as the earth under their feet began to move, as buildings collapsed around them and waters rose to drown them, as fear of death and despair over the loss of everything they owned took hold of them. An account of this kind would move, it would touch the heart and perhaps improve it. I propose to leave this story to others, more skilled in the art. I shall here describe only the work of nature.[22]

Kant emphasizes in particular that earthquakes have natural causes. He suggests that the reader perform an experiment: he should take twenty-five pounds of iron shavings, the same quantity of sulphur, mix the two with water and bury the mass a foot and a half deep in the ground. The soil should be stamped hard. In a few hours vapours will begin to rise out of the ground, the soil will begin to shake and flames will begin to shoot out of it. Only thus, by understanding geological processes can man understand the natural catastrophe which had befallen Lisbon. Besides, Lisbon was not the only city in Portugal to have suffered the calamity; other towns have felt tremors as well and have been assaulted by the unusually high floods. The earthquake was not a punishment visited upon Lisbon but a chronic natural process. The quakes continued throughout November and the succeeding months, at intervals of several days, and their activity was felt in different cities of Europe and Africa. And they were not always destructive: in some places they were beneficial. At the springs of Teplitz, for instance, the water first seemed to dry up and then began to flow with increased force. Kant includes in his booklet a chapter on "The Benefits of Earthquakes." His enlightenment optimism remains unshaken. He retains the Wolffian point of view, although tempered by sober observations: "Man is so preoccupied with himself that he considers himself the sole aim of God's designs, as if they were woven around him alone, and the laws of the universe were there for his sake alone. We know that the whole of nature is an instance of God's wisdom and his forethought. We are part of it, yet we would be the whole."[23]

The booklet ended with the following emotional outburst: "The ruler who moved by noble instincts allows human misery to deliver man from the afflictions of war, even though they are only part of the misfortunes that assail him, is a beneficent instrument in the gentle hand of the Lord, a gift to the nations which cannot appreciate its greatness properly." Now, that is an unusual ending to a scientific treatise!

However, when we remember when the lines were written, then they become more comprehensible — the spring of 1756. Europe breathes war. One does not have to be a prophet to feel the coming of a catastrophe far more bloody than the earthquake of Lisbon. The philosopher calls upon his king to exercise reason.

But Berlin is in the midst of preparations for war. A realignment of powers had taken place in international politics, a veritable "diplomatic revolution." The traditional rivals, France and Austria, have found a common language. They are apprehensive about the growing power of Prussia which has, in the meantime, formed an alliance with England. The Russians, the Saxons, and the Swedes have joined Prussia's enemies. The alignment of powers does not favour Friedich II, but he, heartened by the two earlier conflicts with Austria, initiates hostilites.

Friedrich did not know that the war was going to last seven years, devastate his country and bring it to the brink of ruin. In August of 1756 his armies easily overran Saxony and pressed forward into Austria. Victories were followed by defeats. The French were brilliantly routed at Rossbach, and Friedrich became a national hero (the French had been considered invincible, all the German principalities on the Rhine cowered before them). In the summer of 1757 Russia entered the war. At the end of August the Prussian corps under General Lehwaldt was defeated at Grossjägersdorf in East Prussia. Field Marshal Apraksin hesitated to advance and ordered instead the retreat to Kurland, to winter quarters. It is possible that he had received secret instructions from Petersburg: Empress Elisabeth had fallen ill, and there was a change of course in the offing. The empress recovered, Apraksin was relieved of command, the new commanding general, Fermor, was ordered to advance. Circumstances were favourable: the Prussians had left East Prussia in order to defend Pomerania against the Swedes.

Fermor, at Memel, pushed his army towards Königsberg by the shortest route — along the Nehrung and over the ice of the bay of Kurland. On 22 January the Russians entered the capital of East Prussia. Andrey Bolotov, a participant in the campaign, tells the story:

> Streets, windows and roofs of the city were full of people. There were so many of them because they wanted to see our army and its commander; the bells of the city were pealing, trumpets and drums echoed them from all the forts and campaniles during our entrance, everything combined to make it magnificent and noble.
>
> Count Fermor took his quarters in the royal castle in the same room where Fieldmarshal Lehwaldt had stayed before him; there he was attended by the authorities, the nobility, the high clergy, the merchants, and other prominent citizens of Königsberg. They greeted him, accepted the protection of the empress and as her subjects begged him to maintain order and discipline. That was promised them.[24]

A new administrative unit appeared in the Russian empire. On 24 January Königsberg took oath of allegiance to the empress. The pastor read the German text, the present repeated it after him; then they signed the oath in their own hands. *Privatdozent* Kant took the oath with the faculty of the university. He, like everybody else, undertook to be "loyal and true to the Illustrious and All-powerful Empress of all the Russias, Elisabeth Petrovna, etc. etc., and to her heir, His Imperial Highness the Grand Duke Peter Fyodorovich, and to serve without reservations the highest interests of Her Imperial Majesty; and should anything be undertaken against them not only will I inform the authorities forthwith, but also try to thwart the deed with all the powers at my disposal."[25]

The Russian two-headed eagle replaced the Prussian eagle on the gates of the city and in its offices. In the homes of the wealthy there appeared portraits of Elisabeth Petrovna. Services of thanksgiving were offered in the churches.

In the wars of the eighteenth century the population was almost never affected; the religious fanaticism which had caused so much blood to be spilled in the preceding epoch, had abated; total extermination as a means to victory had not yet come into being. In the course of military campaigns there were, of course, occurrences of murder, rape, plunder and ravage, but man tended to look upon those as natural catastrophies. The bloodletting was the business of kings and their armies. Once a city had surrendered to the enemy its administrators remained at their posts; the victor was satisfied with his tribute, and life went on as before. When the fortunes of war changed and the previous rulers returned, collaboration with the enemy went, as a rule, unpunished. Friedrich's displeasure with the Königsbergers who had offered allegiance to Elisabeth was expressed solely by his refusal ever to visit the city again after the conclusion of the war.

In February a decree arrived from Petersburg, from the empress: all the "privileges, freedoms, concessions and rights" of Königsberg were confirmed; the freedom of religion was guaranteed, as were the freedoms of trade, movement, private property, and so on. A special section of the decree pertained to the University of Königsberg: its budget remained unchanged, the salaries of its teachers unaltered. Instruction was to proceed in full freedom. "Students may continue and complete their studies at the academy, and everything is to continue as before."

General Fermor was appointed the governor of East Prussia. A German source gives the following description of the Russian occupation of Königsberg: Fermor prosecuted all excesses against the established order with utmost severity. Plunderers were executed by firing squads. He and his officers — among whom there were many from the Baltic area — regularly visited the university, official ceremonies in the main auditorium and the lectures of Kant, who was then a *Privatdozent*. Kant gave private lessons to the Russian officers on mathematics, fortifications, strategy and pyrotechnics. The empress of Russia wanted to show her best face, so administration was entrusted to the compassionate and just officers. Fermor introduced novelties: delicacies of the Russian-French cuisine were served at banquets, there were balls and costume parties in which the young Kant took part with pleasure. Königsberg awoke from the sleep of provincialism.

Unfortunately, the testimony here does not come from an eyewitness; the account is based on the oral tradition of the Fermor family,[26] and the tradition is not always trustworthy. The fact is that the count did not stay in Königsberg long. He took his troops to Pomerania and created there a new theatre of military activity. In the summer of 1758 he defeated the Russians in a bloody battle at Zorndorf, near Küstrin. The Russians halted Friedrich II, however, and a year later they inflicted a terrible defeat upon him at Kunersdorf. In the autumn of 1760 they occupied Berlin but not for long. The king was surrounded and fought on all sides, and the outcome was frequently in doubt. But the allies hesitated, intrigued against one another, and the catastrophe was averted.

In the meantime Königsberg lived in peace. The occupied territory enjoyed a privileged status in comparison to the central governments of Russia. Men here were not pressed into service, and the expenditures for the maintenance of the armed forces were not excessive. At first there were disturbances: once, in the middle of the winter, there were rumours that Friedrich's armies were approaching. An order was issued that all the houses were to be supplied with combustibles, so that the city might be set afire in case of a retreat. For a few days citizens walked about with their heads hung low, but the worries were unfounded. When action moved to the Oder, and the Russian army went to its winter cantonments behind the Vistula, East Prussia was so far in the rear that the misery of war was forgotten.

A new governor arrived from Petersburg — Baron Nikolay Andreyevich Korff, a dim-witted and unpredictable magnate whose carrer was the result of his connections at the court. He delegated government to his subordinates and devoted himself to the life of pleasure. He was fantastically rich and used to living on a grand scale. Official-banquets, balls, costume parties followed one upon another. And each time there was in attendance half of the Russian officer corps, the city nobles, and even distinguished prisoners of war. The frenzied extravagance of such affairs horrified the parsimonious citizens of Königsberg. A troupe of actors was brought in from the enemy territory — from Berlin. The costume dances were at first held only for invited guests in the residence of the governor; then they were held at the opera, and the admission was free. "The best musicians travelled to Königsberg from afar," A. Bolotov accounts; "for each new dance we had new musicians and new dances. In short, all that was best and new had to be seen and heard here, and one can say with certainty that since the establishment of their kingdom Prussians had never seen such pomp, entertainment, and pleasure in their capital, and that they will in all likelihood never see it again. I doubt that the kings of Prussia ever lived so merrily, so grandly, and with such abandon as our Korff."[27] Countess Charlotte Keyserling who had been Kant's patroness shone at the governor's balls, Korff was in love with her.

With the governor from Petersburg came a staff for the chancery — two secretaries, a protocol officer, chancery clerks, coypists of various ranks. The staff were carefully selected, but one important detail had been overlooked: an interpreter; the officials did not speak a word of German, while the population — how annoying — did not speak Russian. Korff himself, a Baltic German, spoke the official language of the empire quite poorly. And so they had to seek interpreters among the soldiers. So it happened that the infantry lieutenant Andrey Bolotov (whose memoirs represent not only an historical source but also a literary memento of the times) found himself assigned to the office of the governor. For a long time he was the only man carrying on the actual business of government who was proficient in both languages. All petitions, complaints, and other papers passed through his hands. It is probable that he handled *Magister* Immanuel Kant's petition in this capacity.

The business of the university had not been interrupted. Kant taught his courses and lectured to the Russian officers. Kant taught the subjects of fortifications and pyrotechnics. It is possible that Grigory Orlov, the future favourite of Empress Catherine was among his listeners; he had been wounded at Zorndorf and was convalescing in Königsberg. Perhaps Aleksandr Vasilyevich Suvorov, then a lieutenant-colonel visiting his father, General V.I. Suvorov, in the Prussian capital, was there as well. The latter would relieve Korff as the governor. Kant's friend Sheffner writes in his memoirs that Kant came to know the future Russian generalissimo in those years.

In December of 1758 Professor of Philosophy Kypke died. There were five applicants for the vacated chair. Kant was among them — urged on by his old patron, Pastor Schultz, now a professor of theology and the president of the university. The academic senate narrowed the choice down to two candidates — Buck and Kant. On 14 December 1758 Kant sent a personal petition to Empress Elisabeth, which we here reproduce in entirety:

Most Serene and All-powerful Empress
Autocrat of All Russians.

Gracious Empress and Madam!

With the death of Doctor and Professor Kypke, of blessed memory, the chair of logic and metaphysics in the Academy of Königsberg has fallen vacant. These sciences have always been the special fields of my study.

Since becoming a lecturer at said university, I have taught logic and metaphysics every semester. I have publicly defended two dissertations in these subjects; in addition, I have published four treatises in the learned journal of Königsberg, three *Programmata* and three other philosophical tracts which give some sense of my research.

Encouraged by my hope that I have proven myself worthy of serving the Academy in these sciences and deeply aware that Your Imperial Majesty is a mighty protectress of sciences and their advocate, I most humbly beg Your Imperial Majesty graciously to appoint me to the vacant post of full professor, the more so as the academic senate have seen fit to append to my most abject petition their judgment of my abilites.

I remain Your Imperial Majesty's most devoted and humble servant

Königsberg, 14 December 1758 Immanuel Kant

The text of the petition was first published from a copy in 1893 by the University at Dorpat (today the University of Tartu); it is not known who had made the copy and from which original. Jury Bartenev, who published the Russian translation (in the journal *Russky Arkhiv*, 1896, No. 7) wondered where the original was and who could have submitted it to the empress. The author of this book has searched the archives, unfortunately without success. There is no trace of Kant's petition in either the Archives of Russian Foreign Relations, a repository of material from Königsberg and of imperial decrees pertaining to the kingdom of Prussia, or in the Central Archive of Old Documents, where all the documents of the Königsberg chancellery are kept. (There is a petition there, submitted by Professor I.G. Bock, wherein he asks that his payment of dues be waived; Bock was the holder of a chair in literature who greeted the Russian armies with an ode in honour of Elisabeth; the verses were sent on to the empress who liked them so much that she ordered the author be rewarded from the state treasury and elected to the Petersburg Academy of Sciences; as a member of the Russian Academy, he was exempt from paying dues.) Evidently, the question of vacancy was settled in Königsberg, not in Petersburg.

The decision went against Kant. The professorship went to Buck, who was older and had more years of teaching experience. Perhaps another factor played a role in the decision. Andrey Bolotov, (whom we have already mentioned) occupant of a responsible position in the chancellery, showed a lively interest in philosophy. He found Wolffianism repugnant, it seemed to him to be the seedbed of cynicism, freethinking, and atheism. (Voltaire and Helvetius, Bolotov thought, were "corruptors and debauchers of the human kind.") Reading of Wolff's works brought him to doubt the truth of the dogmas of revelation and he experienced a profound crisis of conscience. A chance purchase of a sermon by Crusius, for a penny, turned out to be his salvation.

> The sermon was not only theological in nature: it was also philosophical. The noble man was able to set out the great importance of a firm faith in the truth of revelations and the dreadful dangers of doubt. As I read it, I shook from head to foot, his words and his convictions wrought such a change in my reason and in my heart that I felt as if a heavy burden had been lifted from my shoulders, the turbulence in my veins ceased, everything became peaceful again. I felt drunk with my own self: when this great man, worthy of the highest respect, so fiercely stands in defense of revelation and so wisely and convincingly speaks of the need to believe in it, how can I doubt. I who am a thousand times lesser than he? No, no! I said to myself, from this moment on this will not happen and I will not fail to follow his advice. In a word — the wonder of the incident struck me down to my knees and I tearfully offered my thanks to the almighty, for his grace was clearly upon me, and I begged him to shed light on my way in the future; and since I had received his assistance I resolved to do what Herr Crusius demanded of his readers and listeners, namely, to read everything that has ever been written in the world in defense of the truth of the law of divine revelation.[28]

Kant was openly anti-Crusian. As for his religious convictions, even the university authorities sympathetic to him were not certain of their orthodox purity. "Do you live in the fear of God?" President Schultz asked him, and as soon as Kant answered in the affirmative Schultz put his name up for the professorship.

Bolotov's memories of those days are detailed; they cast light on the life in Königsberg, but Kant's name does not appear at all. There is frequent mention of his opponent at the university, the Crusianer

Weymann, whose lectures Bolotov attended with zeal as they reinforced his dislike of Wolffianism. Bolotov learned Crusius' texts by heart and translated them into Russian. Weymann was touched and considered Bolotov his best student. It is possible that the newly converted Crusianer gave the chair of philosophy to the mathematician Buck whose indifference to the acute problems of world view was preferable to Kant's Wolffian stand.

In October 1759 Weymann stood for habilitation. His dissertation was entitled *De mundo non optimo*. To claim that our world is the best of all possible worlds is to limit the freedom of divine will, Weymann averred. Kant declined the role of the opponent in the defense, but a day after the event his booklet appeared — *On some Observations about Optimism*, a prospectus of his lectures for the coming winter semester. The booklet contained Kant's polemic with Crusius and his follower, Reinhardt, the laureate of the Berlin Academy of Sciences competition. Weymann's name was not mentioned, but he felt he was at the receiving end of the polemic and retaliated by publishing *An Answer To 'On some Observations about Optimism'*.

Kant abstained from continuing the polemic; he thought that the notion of the perfection of our world was given sound and unambiguous grounding in the prospectus. At first glance the idea appeared self-contradictory: just as it is possible to add a unit to any series, it is possible to add a new reality, a new perfection to any sum of realities. Kant demurred: reality is not an aggregate; the greatest number in reality does not exist, but an absolute reality is not only possible, it exists in God. "Precisely because God chose from among all the possible worlds this one, he must have held it the best, and as his judgment admits of no error this *is* the best."[29]

Many years later Kant would call his condition during the *magister* period "a dogmatic slumber." He forbade the use of his earlier works, and as for the tractat on optimism, he wished all the extant copies to be destroyed. Indeed, it is difficult to understand the following tirade: "When I consider that I have been chosen to be an insignificant link in the most perfect of all possible designs, I, who am nothing in myself and exist only because of the whole, value my being (*Dasein*) all the more highly, because I was elected to be part of the best plan of creation . . . The whole is perfect, it confers perfection to everything."[30] It is difficult to tell what prevails more in the statement — a primitive church dogma or a shallow dogmatism of the Enlightenment. "I, who am nothing in myself. . ." How powerful is the contrast between this degrading declaration and Kant's future motto: "Man is an end in himself."

Everything is good. Everything is for the best. But what shall we say when a youth dies in the bloom of his years? Should his mother not rebel against the heartless almighty? Kant writes a letter to the mother and then publishes the letter as a pamphlet: *Thoughts Upon the Premature Death of Johann Friedrich von Funk* (1760).

The ways of providence, Kant assures the reader, are always and in every respect wise and honourable. The premature death of those upon whom we build our hope strikes terror in us — and yet: how often is that an expression of heavenly mercy? Does not the misfortune of many consist in death's coming too late? Kant's point is that those nearest to Funk should not mourn but rejoice at his demise. "Temptations, arising to crush the fragile virtue, misfortunes and vicissitudes nesting in the bosom of the future, all these were spared the fortunate soul taken away from us by an early death, yet in a blessed hour."[31]

Such were the fruits of Wolffian enlightenment. Soon they will turn bitter in Kant's mouth. The awakening out of "dogmatic slumber" is about to begin.

What you are determines the philosophy you choose.
— FICHTE

CHAPTER TWO

"I am learning to honor man"

In 1762 the Seven Years War drew to a close: Prussia stood at the brink of disaster, Friedrich thought that everything was lost, considered abdication and carried poison on his person. Then — news of salvation arrived from Petersburg: Elisabeth had died. Her nephew, Peter III, became emperor. He was half German, married to a German princess, a supporter and admirer of the Prussian king. Things changed immediately. Russia withdrew from the war, concluded a treaty with Prussia and turned its arms against erstwhile allies. Friedrich, who was ready to make territorial concessions and part with East Prussia, quite unexpectedly regained everything he had lost. The Russian guards put on Prussian uniforms and began to follow Prussian drill. The Russian army was readied to fight for Prussian interests.

The frivolous governor Korff had left Königsberg, as had the severe Suvorov. Lieutenant Andrey Bolotov had also returned home. His parting gift to his teacher of philosophy, Weymann, was a sheepskin fur coat. On 8 July the last Russian governor, Voeykov, issued a proclamation releasing population from their oath of allegiance to the emperor. Russian arms were taken down from the gates of the city and administrative buildings, and the Prussian eagle ascended again.

Celebrations and services of thanksgiving were beginning when suddenly the news of a palace coup arrived from Petersburg: Catherine II had ascended to the throne. Voeykov again assumed the powers of the governor, Russian arms reappeared in Königsberg, Russian sentries reassumed watchkeeping.

Catherine II (née Sophie Auguste, the princess of Anhalt-Zerbst) was the daughter of a Prussian general. She was torn between her sympathies for things German and her desire to establish herself on the Russian throne. A compromise solution was arrived at: the treaty with Friedrich was annulled, but the occupied lands were returned to him. In August East Prussia was finally turned over to the Prussians: Field Marshal Lehwaldt returned to the castle of Königsberg. (Friedrich had earned Catherine's gratitude when he had elected her to the Berlin Academy of Sciences. She was the first woman member and until the end of the last century, the only one.)

1762 was a crucial year in Kant's life. One must bear in mind that the work of Jean Jacques Rousseau was of immense significance for Kant's new research which would ultimately culminate in his *Critique of Pure Reason*. Towards the end of the summer Kant got hold of *Emile*. The book, which had been burned in Catholic France and in Calvinist Switzerland, occupied him so thoroughly that for several days running he missed his daily walk. He did not want to interrupt the reading. The only ornament of his study became the portrait of the citizen of Geneva.

Kant admitted that Rousseau became his "second Newton." If the Königsberg philosopher looked at the endless starry cosmos through the prism of Newton's equations, Rousseau's paradoxes helped him to look into the obscure corners of the human soul. According to Kant, Newton was the first to see order and regularity where before him everyone had seen merely a chaotic mass: Rousseau, on the other hand, discovered the principle of the human nature in the chaos of humanity. Rousseau's books freed Kant from the typical prejudices of a theoretical thinker, they democratized his thinking. "I feel a tremendous thirst for knowledge . . . There was a time when I thought that only knowledge brought honour to mankind; I despised the multitude because they were ignorant. Rousseau set me straight. This blinding preference is disappearing, I am learning to honor man . . ."[32] Kant did not merely change his point of view; he underwent a moral renewal, a revolution of all his life's objectives.

Rousseau gained fame with his treatise *Discours sur les arts et les sciences*, with which he won the competition at the Academy of Sciences at Dijon. The theme given the competitors was formulated as follows: "Si le rétablissement des Sciences et des Arts a contribué à épurer les moeurs." *Rétablissement* could be taken to mean renaissance as well, but Rousseau considered progress in general in all of its contradictions. He concluded that progress in the sciences and arts contributed nothing to our true well-being, it only corrupted our morals.

Before anybody else and with more clarity Rousseau expressed the nascent understanding of a time which would reexamine the postulates of the Enlightenment — faith in the endless power of rational thought and the judicious reason of rulers.

The Seven Years War had taught a few lessons. But in Germany ten years would pass before "Storm and Stress" turned a new page in the spiritual life of the country. At first only the best, the most acute minds contemplated the crisis.

Before Kant's eyes there unfolded a remarkable spiritual metamorphosis. In 1756 he met Johann Georg Hamann, a Königsberger and graduate of the university of Königsberg. Hamann was passing at the time through the city of his birth on his way to London, as the representative of the merchant firm Berens from Riga. Hamann was interested in commerce, he wanted to study economics, and he thought that in London he could satisfy both his desires. In London, however, he became interested in other things: the curious, impressionable, unstable youngster gave himself over to the life of dissipation and soon squandered all his money. Illness accompanied penury. In his search for a way out of the predicament Hamann turned to the Bible and discovered in it the long lost support of life. A changed man returned home. The firm was willing to forgive his losses; but together with Kant they attempted to return their friend to the fold of the Enlightenment. In vain — Hamann had something else in mind: with the fanaticism of a proselytizer he engrossed himself in the Testaments, read them in the original, learned Greek and Arabic. Nor did he forget the contemporary literature — English, French, and his native German literature. Soon he began to write.

Magister Kant, "the little *magister*" (as Hamann called him) was a man of different interests, a different cast of mind and character, yet sufficiently broad of outlook to recognize in Hamann an extraordinary personality. Their acquaintance grew firmer. In 1759 Hamann, who lived in Königsberg, wrote a long letter to his friends — an answer to their entreaties to reconsider things. It appears to be a first attempt at writing, a draft of a literary work. Soon a thin booklet came out, entitled *Socratic Curiosities*[33] and dedicated to "None and Two," "None" was the reading public, "two" were Kant and Berens, the merchant from Riga. Hamann mentioned no names but gave striking characterizations. One of them, he said, envied Newton and wanted to become the warden of philosophy (Newton had been appointed warden of the mint by the king). But the monetary affairs of Germany were far more orderly than her textbooks of metaphysics. The wise men had not yet invented a stone to test the truth of their ideas as it was possible to test the content of a coin for noble metals. All this was doubtlessly directed at Kant.

Hamann's style is heavy, full of innuendos and allusions. When he speaks of the sage of Athens he means himself, and his search for truth. In the letter to Kant, Hamann calls him Socrates, Berens Alcibiades, and he, Hamann, is Socrates' *daimon*, i.e., his intuition. In *Socratic Curiosities* Kant and Berens are sophists, proponents of casuistic learning — Hamann is the Christian Socrates. Hamann rejects the postulates of the Enlightenment, just as his great predecessor had rejected the enlightened men of Athens. We think too abstractly, that is the crux of his argument. Our logic forbids contradictions, but truth is to be found in contradictions. To deny contradiction is to kill thought. The Delphic oracle had called Socrates the wisest of men because he admitted that he knew nothing. Who lied — Socrates or the oracle? Both spoke the truth.

Knowledge of the self is Hamann's cardinal concern; in it, according to him, reason has no power, knowledge is only a hindrance; only faith springing from an inner feeling can help. Under his pen Socrates becomes an irrationalist, a precursor of Christianity: the Athenian philosopher wanted to lead his fellow citizens out of the labyrinths of learned sophistry to truth, the truth that lies in "the hidden," in adoration of an "unknown god."[33] That is how Hamann sees his destiny. Kant could not warm to the perspective of replacing one labyrinth with another. He told Hamann as much, and their relations became strained.

Nevertheless, the arguments of the "Magus of the North" provoked thought. Kant could not but give deep thought to the new book by Hamann, *A Philologist's Crusades*, a collection of essays whose central piece carried a somewhat unusual title — "Aesthetica in nuce." The term aesthetics was particularly unusual. It had been reintroduced not long before by Alexander Baumgarten to designate the doctrine of the beautiful, and to Baumgarten it meant the theory of sensual knowledge. To him, a follower of Leibniz and Wolff, aesthetic knowledge was a lower order of knowledge. On the first page of his essay Hamann categorically states that the opposite is the case: all the wealth of human knowledge is contained in sensual images, there is nothing higher than the image. The image is wholeness — "parts are to be rejected."

"A brilliant maxim," commented Goethe in *Poetry and Truth*, "but it is difficult to follow it! It can be applied to life and art, but there is a difficulty in every transaction entailing the word which is not poetically charged: in order to say something, to convey a meaning, the word must free itself, it must become a part. When man speaks he must become one-sided, if only for a moment; there is no communication, no knowledge without division."[34] Kant could have said the same thing. And yet — the critique of abstract thinking, of Wolffian one-sided systematics must have attracted Kant's attention.

Hamann blamed the Wolffians for having turned scholastic, thereby abandoning life, nature. They wanted to become rulers of nature, yet they bound themselves by the hands and feet. Thus masters became slaves. Then came from the other end of Europe another invective which did not deny merely a school of philosophy but the whole of contemporary civilisation: Kant did not read it with indifference. Hamann prepared Kant for Rousseau.

Kant did not become a Rousseauian; from his studies of geography he had a fairly extensive knowledge of distant peoples and he was not prepared to idealize them. "Rousseau. His approach is synthetic, he begins with the natural man. My approach is analytical, I begin with the civilised man . . . By nature we are not saints . . . The Arcadian pastoral and our gallant courtly life are equally *passé* and unnatural, although alluring. Pleasure cannot be practised like a profession."[35]

These notes from the sixties show that Kant's attitude towards his favourite is critical.

In addition to Rousseau Kant named David Hume as the thinker who helped him awaken from his "dogmatic slumbers." The enthusiastic Frenchman and the sceptical Englishman, two opposites, were fused together in Kant's contradictory nature. Rousseau "straightened out" Kant the man as a human being and moralist; Hume exerted an influence on his epistemological questions and forced him to reconsider metaphysical dogmas.

On the eve of the winter semester of 1762 Kant, as was his habit, published a booklet — an announcement of his lectures. Before, he had considered problems in the natural sciences. Now, the theme was a philosophical one with the title *On the Proof of the False Subtlety of the Four Syllogistic Figures*; it was a first, still timid, but promising attempt at a critique of the formal logic which was grounded in the support of Wolffianism. Kant calls formal logic "a giant with feet of clay"; although he takes a swing at the giant, he does not hope to topple him.

For Kant logic must make good the claim of investigating the genesis and formation of concepts. Concepts proceed from judgments. But where is that mysterious power which makes judgment possible? Kant's answer: judgments are possible thanks to the ability to convert sensible impressions into an object of thought. This is a significant answer: it shows Kant's first, as yet limited, endeavour to construct a new theory of knowledge. Until now he had shared the Wolffian admiration of deduction, he was certain that there was no limit to man's ability to develop new concepts out of the ones he already had (although his research in the natural sciences rested on experimental data). Now he wonders how one can introduce experimental knowledge into philosophy.

This pressing problem dominates another work, written towards the end of 1762 and published two years later under the title *An Enquiry into the Clarity of the Principles of Natural Theology and Morality*. The work was written in connection with a competition announced by the Berlin Academy of Sciences. The purpose of the competition was to obtain answers to the question whether philosophical truths, in particular the principles of theology and morality, were susceptible to proofs of evidence, like truths of geometry; if they were not, what was the nature of the proof of these principles, to what degree were they certain or were they somehow totally convincing?

Learned men from other countries were invited to take part in the competition (except for the members of the Berlin Academy, who were going to judge submissions). The prize was a gold medal worth fifty ducats. Clean and legible manuscripts were to be sent to the permanent secretary of the Academy, Formey, by 31 December 1762. Authors were asked to put their names in an accompanying sealed envelope, which should bear the same motto as was placed on the manuscript.

The theme fitted Kant perfectly: he was just waking up from his "dogmatic slumbers" in the arms of Wolffian metaphysics. He compared philosophy to mathematics and dealt with the qualitative multiplicity of objects in the former as compared to the latter. Compare the concept of a trillion with the concept of freedom! The relationship between a trillion and one is clear to everyone; but no one has yet succeeded in making the concept of freedom comprehensible, i.e., no one has been able to establish a relationship between this concept and the multiplicity of simple and known concepts out of which it is composed. Naturally, many consider philosophy a lesser science than mathematics, but they consider philosophy everything that is contained in books described as philosophical. The point is that there is *no* book of original philosophy. Philosophy must adopt the method which Newton had introduced into the natural sciences, which had such fruitful results there. Man must rely on experimental evidence and seek universal laws.

But, does this hold true for theology? With what experience can one prove the existence of God? The experiential evidence of philosophy is not only the evidence of senses, but also the "inner experience," the unmediated, evident consciousness. Thanks to the latter, the knowledge of God is certain.

The submission for the competition required also that an answer be provided to the question regarding the principles of morality. Here, in Kant's opinion, matters stood much worse than in theology; no level of necessary certainty had been reached. But here too a compulsory grounding of morality is possible. And then Kant articulates a position which would be of great import in his future philosophical development: man must not confuse truth and happiness, knowledge and moral feeling.

Another significant position is stated in the treatise, almost in passing: it has to do with the role of unconscious representations. The problem of the subconscious entered into science in the seventeenth century. Locke rejected the possibility of existence of unverifiable psychic activity.

If one believes that the soul thinks and man is not aware of its thinking then we have split man in two. If man thinks in his sleep and does not know it, then the sleeping man and the waking man are two different persons. The sleeping Socrates and the waking Socrates, Locke believes, are not one and the same person: Kant, ever more involved with dialectical procedures, could not be confused by such distinctions. He would later divide consciousness in two spheres: he would not find it difficult to find the other pole of consciousness, the pole which is different from it but not absolutely exclusive of it. Leibniz had already decisively turned against Locke; he had discredited as a source of dangerous error the notion that our soul has at its disposal only those perceptions of which it is consciously aware. The nonconscious material Leibniz called "petites perceptions;" if perceptions are small, their significance is great, because they form habits and taste. Kant introduces another term: "dark representations." He who negates their importance "passes carelessly by a great secret of nature, *viz.*, that it is quite possible that in deepest sleep the soul is capable of rational thought."[36] We should remember this: here is a key to one of the most important sections of the *Critique of Pure Reason.*

Kant completed the work in haste; time was short. As a result, parts were left without sufficient argumentation, there was often no smooth transition from one part to the next. Kant was aware of the weaknesses of his work; in an addendum he promised to make all the necessary improvements should the treatise be received with favour.

It took the judges half a year to arrive at a decision. Then Kant read in a Berlin newspaper that the first prize had been awarded at the meeting of the Academy on 31 May 1763. The official announcement stated that upon opening of sealed envelopes it was determined that the winner of the competition was as the official announcement read, "the Jew Moses Mendelssohn, a local resident." At the same time the Academy made it known that a German work, under the following motto "Verum animo statis haec vestigia parva sagaci sunt, per quae possis cognoscere caetera tuto," was deemed nearly equal to the work of "the learned Jew." Kant had used the distich from Lucrecius' *De rerum natura* as the cypher for his work.

The award, then, had gone to another, but what did the jury mean when they characterised his work as being almost equal to the work of the winner? Kant wrote to Professor Formey immediately and asked for a clarification. Did it mean that his work was going to be printed together with the winning entry? If that was the case, could he amend and correct the text? The answer from Berlin was affirmative: both works were going to be published together and their authors could do the necessary revisions. Kant, however, found no time to revise the text, which was published in its original form.

There was another pleasant surprise from Berlin: a positive review of his conspectus *On the Proof of the False Subtlety of the Four Syllogistic Figures*. The anonymous reviewer (it was Mendelssohn, the winner) called the author a daring man who was going to shake German academies to the core.

A future revolution in philosophy was also announced by the ideas contained in Kant's *An Introduction to the Concept of Negative Quantities*. Kant admitted regretfully that the problems he wished to consider were not yet quite clear to him; nevertheless he wanted to publish his work because he was firmly convinced that those quantities were important and that even imperfect philosophical efforts could be of use — perhaps someone else would provide answers to the question he had raised.

Kant is interested in the problem of the unity of opposites. The starting point of the consideration was stated in his dissertation, namely the difference between the real and the logical ground. What is logically correct does not have to be true in reality. The logical contradiction consists in simultaneously affirming and denying something about the thing. Logic forbids accepting both predicates as true at the same time. Man cannot claim that his body is simultaneously in motion and at rest. One statement negates the other, and the result is nothing.

A real contradiction, entailing opposing forces, is something else. Here one annuls the other, only the result is not nothing, but something. Two equal forces can act upon a body in opposite directions: the result is that the body is at rest, and that is also an existent fact.

The world around us is full of such real contradictions. The mathematical doctrine of negative quantities has been operating with the concept of real contradiction for a long time. Philosophy must borrow certain principles from mathematics, the truth of which principles is demonstrated in nature. This is particularly the case with the real contradiction because it can be observed not only in nature but also in man's behaviour. Pleasure and pain stand in the relationship of positive and negative quantities. Kant illustrates his notion with an example. When a Spartan mother learns of the heroic feats of her son her soul is filled with pleasure. But when she learns that her son has fallen on the field of honour, her pleasure is diminished. Kant suggests that the degree of pleasure caused by the first news be designated as 4a; if we presuppose that the pain of the second news is simply 0, when we add the two numbers together, thus, $4a + 0 = 4a$, we are forced to conclude that pleasure was not diminished by the death notice, which is improbable. If we designate the pain with a negative quantity, for instance, $-a$, we obtain the correct result: $4a - a = 3a$.

The fact that Kant translates feelings into numbers — and the result of the operation — does not speak well of him. Yet, we should not judge the philosopher too harshly: we know that he is learning to "honor man." Learning takes time, it is not easy to abandon old prejudices.

In the winter semester of 1762 Kant got a new student, whom he noticed immediately. The young man was enrolled in the school of theology, was unusually gifted and assiduous, and wrote fairly good verse in which he imitated Kant's favourite poets — Haller and Pope. The youth set one of his teacher's lectures in verse and handed it to the teacher at the next meeting of the class; Kant liked it so much he read it from the lectern. The student had already written an ode on the occasion of Peter III's ascension to the throne. The student was poor, so Kant did not charge him for instruction.

Johann Gottfried Herder, the son of a sexton and teacher in the parochial school at Mohrungen, had never thought of higher education. Then chance took a turn. A Russian regiment was quartered for the winter in his town. The regimental doctor, Schwartz-Erla, took an interest in the fate of the well-read youth, who had helped him with Latin translations. He took Johann Gottfried with him to Königsberg, determined to make a physician out of him. At the first anatomical section the sensitive youth fainted; Herder had to change to theology.

Herder took all of Kant's courses — metaphysics, ethics, logic, mathematics, physical geography. He took careful notes and at home put them in proper order. All his notes have been preserved: they are clear, accurate, they develop further the problems Kant had touched upon.[37] The lecturer, for instance, puts forth the notion that soul is a simple substance. Good, but does that mean that it occupies space? If it does, then the soul is something material and could be measured. Can one imagine one cubic inch of soul? And who can tell me where the soul is situated in the human body? Does that mean that spirits have no bodies? Can they, by any chance, be possessed of special organic bodies? How else can they be and act in the universe? Magnetic force also exists, invisible and material at the same time. There is only one conclusion to be drawn: the soul has an inner nature, known to us by the fact of consciousness; as for its nature — we know nothing.

Or another problem: does the soul survive the death of its body? Quite likely. (Note that the lecturer of the University of Königsberg refuses to give a definitely affirmative answer to one of the cardinal questions of Christian teachings. Lieutenant Bolotov was right in his wary view that Wolffianism elevated doubt to the status of principle.) What are the arguments in favour of immortality? If I must await my complete annihilation then my being is nothing but a toy in the hands of my maker. My reason protests and I would then prefer not to live at all. I say that the world is not a pile of rubble but a certain unity, a whole; if that is so then there must exist a chain, a link between the past and the future. Thus we arrive at the question whether man is that link. Man enters upon life as the result of a chance concatenation of circumstances. Conception is the product of chance, as is the survival of the embryo and the newborn infant. Some live so short a time that they cannot establish the sought for connexion between the past and the future. But what if one lives long? Each life is short in comparison with the infinite task of learning. The learned man can in his old age revert to childhood. Newton knew no pleasure, no respite, no rest, he lived only his science and then ended his life as a childish, funny old man. Is it not better to live without a thought for tomorrow, in the circle of one's friends? Go, seek pleasure! Death awaits. (Kant seems almost to be thinking aloud, he seems to have reached a fork in the road. There is nothing left of the confidence of the youth who was proudly certain that he had "chosen his way.")

Kant speaks about god. We cannot find proof of his being in either our own experience, or in the experience of others. We can only rely on reason: only a system of rational rules can bring us to the conclusion that there is a higher, absolute and necessary being. (Kant laid out his considerations of the theme in *The Only Possible Ground of Demonstration of God's Being*. The work came out in 1763 and brought its author his first literary recognition, but theologians demurred. *Magister* Weymann soon published a rebuttal; in the Catholic Vienna the book was placed on the index of forbidden books.)

There remains yet another question: is such a view of religion not destructive to the foundations of ethics? Kant follows Bayle and Hutcheson and affirms that religion and ethics are different things. Ethics is a matter of human rather than divine rules. Of course, a god without morality is a terrible thing, but there are such instances (to a Hottentot the Christian god bears the visage of a Dutch captain). Morality can exist independently of religion. There are peoples who know no god. Society should tolerate atheists as long as they behave morally. Spinoza was an honourable man. In education the first task is to awaken moral feeling and then implant the concept of godhood, otherwise religion turns into a prejudice and man into a liar and pharisee. Man must first develop an inner sense of duty and then extend that sense to his fellows. An education leading to moral feeling must precede one leading to obedience. Act in accord with your moral nature — that must be the fundamental law of behaviour.

Giving a definition of the moral nature of man is, naturally, the major difficulty. If a Spartan woman was cast in the street without her clothes, that was for her a fate worse than death. But in Jamaica women go about naked. It is forbidden to marry one's sister; in ancient Egypt such marriages were regarded as sacramental. Eskimos, who kill their aged parents, in actuality perform an act of mercy because they spare them a lingering death on a hunt. Rousseau was right in giving thought to what was natural and what was artificial in man; he was absolutely right in concluding that the one-sided development of sicence brings misery in its train.

Herder left us not only a vivid picture of Kant's mind at work; he also gave us a sharply etched literary portrait of his teacher. The latter he drew in later years, when their relations had become inimical, so there is no need to fear that he was trying to flatter his master:

> I remember my youthful friendship with the philosopher with gratitude and pleasure. He truly taught me what man was. In the bloom of his life he was full of youthful ardour, and I believe that it will not desert him in venerable old age. His open visage, made for thought, was the seat of serenity, from his eloquent mouth flowed pleasant speech, rich in thought. A joke, a witty remark, humour, he used them always to good effect and at the right moment, remaining serious amid the general hilarity. His lectures were fascinating discourses; he would speak of an author

> and one could see his mind at work, probing further and further, and yet, in the three years I attended his lectures daily I never noticed in him the smallest sign of arrogance. He had an opponent, bent on proving him wrong, and he never paid him attention . . . I heard his evaluations of Leibniz, Newton, Wolff, Crusius, Baumgarten, Helvetius, Hume, Rousseau, some of them then very new names; it is remarkable that when he dealt with them his sole aim was a noble search for truth, a beautiful enthusiasm for everything that was best in man, a ceaseless, dispassionate desire to imitate what was best and greatest. He did not know what intrigue was, sectarianism and prejudice were alien to him, he did not try to enlist followers nor did he strain to have his name on the lips of the young. His philosophy provoked independent thought, and I can think of no more efficient agency to this end than his lectures, his thought was born before your eyes and then you had to develop it further: he had no patience with dictation, dogmatic pronouncements, lecturing. Natural history and physics, the history of man and peoples, mathematics and experience were to him sources of human wisdom, from them he drew life-giving succor. To them he referred his listeners; his soul lived in society . . .[38]

All who knew Kant say that he was a companionable, responsive man. He worked much, he loved his work, but he knew much else besides. He could relax and let go, temper his deep learning with worldly charm.

"Blessed is he who was young in his youth . . ." (A. Pushkin) After his lectures *Magister* Kant spent his time over a cup of coffee or a glass of wine; he played billiards and in the evening, cards. Often he made his way home after midnight, once — on his own admission — so drunk that he could not find his way to his street where he lived in the sixties. He had to get up early: the classroom awaited. His weak health demanded a more regular regime.

The physical weakness, which had been his lot from childhood, was joined with the passage of years by a spiritual ailment. Kant called it hypochondria. The philosopher described the symptoms of that debilitation in one of his works: "The trouble with hypochondria is that wherever the pain may be, the nervous plexi in all parts of the body respond to it. A melancholy fog envelops the soul, causing the patient to come down with whatever illness he may have heard of. Therefore he likes nothing better than to talk about his illness, he devours medical books and finds symptoms of his illness everywhere."[39] Company exerts a beneficial influence upon the hypochondriac: his spirits and his appetite improve in it. That may be why Kant never ate alone and in general loved company.

He was much in demand, and he never refused an invitation. As a clever and lively conversationalist Kant was the life of the party. He was natural, easygoing, quick-witted. Once, at a dinner, a young lieutenant spilled red wine in the presence of a senior officer and was ready to fall through the floor with embarrassment. *Magister* Kant, who was talking with the senior officer about a battle, calmly spilled some of his wine on the tablecloth and began to draw red lines to represent movements of troops. Just as he had recently taught Russian officers he now gave private lessons to Prussian officers; a general's carriage was sent to fetch him, two generals were among his good acquaintances.

About friendship Kant had a favourite saying: "Dear friends, there is no friend!" This Aristotelian utterance was meant as a joke, but in every joke there is a grain of truth, as we know. The truth is that Kant thought of friendship very highly (he valued it more highly than love because friendship includes love but requires respect as well); he was inwardly ready for friendship, sought it, believed for a while that he had found it, but actually there was never anyone about him capable of sharing all of his intellectual interests.

Kant thought about what draws men to one another: a choleric temperament hinders friendship; a sanguine person believes that all men are his friends (he, however, is no one's friend); a melancholy person has few, but good, friends. Kant felt himself to be melancholic.

For this reason one must read Kant's characterization of the melancholy temperament with care. To a certain extent it is his self-portrait. One must not believe, Kant says, that the melancholy man is denied the joys of life and constantly abides in a dark sadness. Not at all; he merely falls prey to such disposition more readily than others under the impulsion of outer or inner influences. Such a person is particularly sensitive to the feeling of the sublime. His pleasures are serious, but not for that reason lesser. He is benevolent, steadfast, and reacts sharply to injustice. That is why he pays little attention to the opinions of others: he relies on his own insight. Such a man keeps secrets well, his own and those of others; he hates lie and hypocrisy. "He has a profound feeling of human dignity. He knows his own worth and considers man as a creature worthy of respect. He will not stand for demeaning servility, and freedom is the breath of his life. All chains — be it the golden chains of courtiers or the iron chains of the galley slaves — are repugnant to him. He is a severe judge of his own deeds and the deeds of others, and finds frequent cause to be dissatisfied with himself and the world."[40]

If the character of such a person changes — Kant also sees this as a possibility — then seriousness turns to gloom, devotion to fanaticism, love of freedom to enthusiasm. Insults and injustices make him vengeful. He then becomes fearsome. He glories in danger and disdains death. If his feelings become changed he is prone to exaltation, he receives signs and omens. He is in danger of becoming a "character," a dreamer or a fanatic. One must know one's weaknesses in order to be on guard against them!

True friendship favours the melancholy character, because friendship and melancholy are equally sublime. Even the melancholy man can lose a fickle friend, but the latter will not lose the former so quickly. The very thought of a past friendship is holy to the melancholic.

Who earned the honour of Kant's friendship? Among his university colleagues Kant had no peers. He considered Joseph Green, an English merchant residing in Königsberg, a friend. In his free time Kant did not like to talk shop, so the company of this well-read man who was innocent of philosophy suited him well. Green and another close acquaintance, the Englishman Motherby, talked to Kant about Great Britain whence, or so Kant thought, came one of his forefathers.

Green, the man of affairs, taught his learned friend the virtues of punctuality. Kant who was not so pedantic in his youth as in his later years acquired the habit late. They tell that once the two men were to go for a carriage ride, for which they were to meet at the gates of the city at eight o'clock. Green was ready at a quarter to eight. Five minutes before the hour he took his hat, picked up his walking stick and set off. When the hour struck, he took his place in the coach and left. He met a breathless Kant on the bridge across the river Pregel, paid no attention to his shouts and drove by. Soon the *magister* became a model of punctuality. He spent every evening at the Green's and left at seven on the dot. One could set one's watch when Kant left Green's house.

Kant also liked to visit Wobser the forester, in whose little house at Moditten, on the edge of Königsberg, he spent his vacations. This was a good place for working as well as resting.

Women did not play the role in Kant's life that they occupied in the life and creation of his younger contemporary, Goethe. He remained a bachelor. But that was not unusual in a philosopher: Plato, Descartes, Hobbes, Locke, Leibniz, Hume were ignorant of marital bonds. Psychoanalysts explain Kant's bachelorhood by postulating that a strong mother cult prevented him from developing attachments to women. The philosopher had a different view: "When I was in need of a woman I would not have been able to feed her: and when I was able to feed a woman I didn't need one anymore."[41] When this admission is coupled with another — "Man can have no pleasure in life without a woman, and she cannot satisfy her needs without a man"[42], then it becomes clear that bachelorhood was forced upon him by the circumstances, and that it brought him no pleasure in his mature years.

"The purism of the cynic and an anchorite's denial of the flesh contribute nothing to the general well-being of the society; they are perverted forms of virtue, unattractive in themselves; devoid of grace they cannot claim to be human."[43] Thus Kant in his old age; when he was in his prime, asceticism was even less appealing. Sexual attraction is according to Kant "a great sensual pleasure," "a special kind of pleasure" that has "actually nothing in common with moral love." Such conclusions are not the result of thought alone.

A certain Luise Rebecca Fritz claimed categorically at the end of her life that the philosopher Kant had been in love with her. According to the calculations of biographers, that would have been in the sixties. Borowski, who was privy to a great deal of Kant's life, confirms that his teacher had been in love twice and had twice contemplated matrimony.

One cannot say that the fair sex was indifferent to Kant either. Quite the contrary. Here, for instance, is an intriguing piece of evidence, a letter Kant received on 12 June 1762. "Esteemed friend! Are you not surprised that I decided to write to you, the great philosopher? I hoped to see you in my garden last evening, but my friend and I searched all the paths and found not our friend under this heaven, so I busied myself with the finishing of a sword tassel — a gift for you. I shall expect the pleasure of your company tomorrow in the afternoon. I can already hear you: all right, all right, I'll come; well, we shall wait for you, my watch is already ticking away. Please forgive this reminder. My friend and I send you a kiss *per simpatie* — the air in Kneiphoff, where you are, is the same and so our kiss will not lose its power of sympathy. Keep well and happy."

The letter was written by a well-known beauty, Maria Charlotta Jacobi. She was twenty-three and married for ten years. The early marriage was unhappy and later fell apart. But what is the meaning of the enigmatic phrase — "my watch is already ticking away"? Quite possibly it is a frivolous jest, a veiled hint at an intimate relationship. One remembers how the gentleman Tristram Shandy, the hero of Laurence Sterne's novel, came into this world: his father was in the habit of winding the great grandfather clock on Sunday evenings and then fulfilling his marital duties; consequently Mrs. Shandy — all in accord with Locke's teachings — became prey to a steady association of ideas "which in truth had nothing to do with each other" — the winding of the clock and sexual congress.

Tristram Shandy came out in 1760 and its fame spread beyond England. Kant loved Sterne. Madame Jacobi was a well-read woman.

Kant was short (157 centimeters, 5′ 2½″) and frail of stature. Tailor's and wig-maker's skills helped hide the deficiencies of his appearance; very fair hair, lively and intelligent blue eyes, a high forehead and good carriage made him quite attractive to the eye. He dressed with taste, in the fashion of his day. (Kant was contemptuous of fashion, called it a matter of vanity, but he also said: "It is better to be a fashionable fool than a fool out of fashion.")

This is how he dressed: a tricorn and powdered wig, a brown coat with black trim, gold piping and silk-covered buttons, waistcoat and trousers of the same colour, a white, ruffled shirt, grey silk stockings, shoes with silver buckles, a short sword at his side. His contemporaries remember him not only as the "little *magister*," but also as the "gallant *magister*." Nothing human was alien to him; when he wrote of man he relied on data he had gotten not just from books.

We must bear this in mind when we read *On the Feeling of the Beautiful and the Sublime*, which went through eight printings in Kant's lifetime and enjoyed great popularity. We have here a new literary enterprise. The philosopher approaches a new genre — the essay. The elevated pathos of early works is gone, humour and irony have appeared, the style is elegant and aphoristic. Kant writes about the world of human feelings and classifies them according to two categories — the beautiful and the sublime. Yet, the treatise is actually not about aesthetics; there are no rigorous definitions in it, everything is only suggested, richly metaphorical, amusing.

The night is sublime, Kant reasons, the day beautiful. The sublime is moving, the beautiful, exciting. The sublime must always be grand, the beautiful can be small in scale.The beauty of a treatment lies above all in the fact that it is accomplished lightly, effortlessly; the overcoming of difficulties excites wonder and belongs to the sublime. The mind of woman is beautiful, the mind of man deep, and by "deep" he means sublime. Woman shun evil not because it is wrong but because it is ugly. Woman tolerates no "ought," no compulsion, no obligation; she does what she does merely because she finds it pleasing. The fair sex is not guided by principles. That is why providence planted the feelings of goodness and kindness in the hearts of women; it also gave them a delicate sense for what is appropriate and graceful; one must not expect sacrifice and restraint of them. In an English paper Kant read that the greatest insult one can offer a man is to call him a liar; the greatest affront to a woman is to be considered deficient in virtue. He, however, thought differently: nothing offends man more than being called a fool, nothing arouses a woman's ire more than being called repulsive.

Man and woman complement each other. In a marriage they develop a single moral personality ruled by the husband's judgment and the wife's taste. It is bad when marriage becomes a unit without harmony, without mutual equality. The fineness and tenderness of feeling develop their full strength in the beginning; domesticity dulls them and then they gradually turn to a companionable love. The preservation of original feelings is a great art, so that indifference and boredom may not destroy the value of the joy which was the sole cause of such a union.

In this treatise Kant sets forth several considerations on the different human temperaments. Again he does not stick completely to his theme; the beautiful and the sublime are only points of departure to which he can add a series of interesting observations. The melancholy temperament abides in the sphere of the sublime. Kant exalts it above others, but he also sees its weaknesses.

The feeling of the beautiful rules the soul of the sanguine man. He loves mirth, the noise of life, change. The joy of others gives him genuine pleasure, but his moral sentiments are devoid of principles and therefore unstable. If a misfortune befalls you he will be sincerely sympathetic, but then slips away quietly until things change. He is generous and given to philanthropy, but forgets to pay his debts. When his character changes he turns vulgar, petty, childish.

The choleric temperament seems to live in the feeling of the sublime, but in fact he is attracted by the false, superficial glitter which leads into error. He is indifferent to what the depths conceal; true goodness does not burn in him and he is just as happy if people only think he is good. His conduct is unnatural; what matters to him is not what he is but what he appears to be. He is always full of himself, whether as a lover or as a friend. In religion he is a hypocrite, in his relations with others a flatterer, in politics a fickle ally. He bends his neck before the mighty of the world and steps on the necks of those beneath him.

The last part of the treatise is dedicated to the peculiarities of national character. This is one of the first attempts at social psychology—a science which only in our day obtained a stronger empirical basis. Kant is satisfied with his own observations here; later, he returns to them time and again: whenever he teaches anthropology. His observations are not always accurate, they are frequently dubious, but they are for the most part very original.

The Spaniard is serious, reticent and upright. His soul is proud (the simplest peasant is aware of his own dignity when he stands before the authorities); he is given to performing deeds of greatness rather than beauty. Even in his personal life he conducts himself with hauteur and a fierce pride. When he spies a stranger he will leave his plough and strut about in his cloak, hand on the hilt of his long sword until the stranger has passed out of sight. He prides himself on not having to work. A Spaniard in love is a curious spectacle: at the bullfight he greets his beloved and in her honour does battle with the dangerous animal. He is fierce: the pyres of the inquisition attest to that.

The Frenchman is in thrall of the morally beautiful. He is polite, courtly, pleasant, somewhat rash in establishing intimacy. He values wit over everything else. He is a peaceful citizen and responds to social pressures only by satire. (That was said twenty-five years before the great revolution; after the revolution Kant discovered in the French national character an "infectious spirit of freedom which engages the reason itself; in the relationship between the people and the state this spirit calls forth a crushing enthusiasm which goes beyond all established boundaries.")[44]

The Englishman disdains everything foreign. For his countrymen the Englishman has established imposing institutions of charity, such as no other nation possesses; but the foreigner who happens to find himself on English soil and in need will be left to die a miserable death. The Englishman is not concerned about appearing intelligent or displaying good manners; he holds himself aloof and even in his own home prefers to eat by himself because a common meal requires a minimum of civility. He is judicious and dignified. He is steadfast, sometimes to the point of obstinacy, daring and resolute to the point of madness. He easily becomes an eccentric, not because he is vain but because he pays little heed to anyone else. (This is obviously a portrait of Green, the merchant.)

The German is a felicitous combination of the feelings of the sublime and the beautiful. He is methodical in everything, including love. German virtues are honesty, diligence, and love of order. Kant considered himself a German. He was proud of that, but without a shadow of complacency and arrogance. (In the *Anthropology* he wrote about the "two most civilized nations," and then added: "It is clear that we are not taking the Germans into consideration here: if the author were to praise Germans he would be praising himself, because he is a German" — the French and the English were being discussed.)[45]

Perhaps, looking into his own hidden depths Kant saw that Germans were prepared to live under any despotic regime. The German submits more easily and decisively than anyone else to the government in power; it never enters his head to rise against the powers that be, or to think of a new order. The Germans are passionate about classification and subordination; they are tireless in perfecting lists of ranks, they are servile out of sheer pedantry because they have invented a whole ladder of ranks and titles: every rung on the ladder corresponds precisely to a determined measure of social authority.

The Arabs are the Spaniards of the East, the Persians its Frenchmen, and the Japanese its Englishmen. For the Indians and the Chinese Kant can find no European equivalents. Of negroes Kant speaks with scorn and calls them to task particularly for their treatment of their wives. "A coward is always a strict master of the weaker; a tyrant in the kitchen fears to be noticed outside his house."[46] Kant writes with admiration about the American Indian. In nobility of character no primitive people are equal to the North American aborigines. Their sense of honour is powerfully developed. Their respect for women surpasses anything seen in our educated part of the world. Women rule there, but they have bought this advantage at a high price: they bear the burden of householding and take part in all the labour performed by men.

In these lucid, occasionally arbitrary passages, there is a deep meaning: they are harbingers of change in the spiritual climate of the land, a gradual turning away from reason to feeling, a lively interest in the unique qualities of a person. Here, as in Hamann, one feels the approach of "Storm and Stress." Kant is ahead of his time. Man has stepped into the circle of his philosophical interests.

The most pregnant passages of the tract did not appear in the original text; they remained as fragments and were published only many years after the philosopher's death. We should note here that from his earliest youth Kant was in the habit of jotting down every thought as it occurred to him. Sometimes he did that on especially prepared notepaper, more frequently he grabbed the first piece of paper that was to hand: an old letter, a merchant's bill, anything. Often the notes are without any scientific or literary merit. But sometimes they are observations of shattering profundity which, one could say, outstrip systematic thought. They are scattered among incomplete phrases as well as polished aphorisms and plans of future works. These reflexions are an extremely important supplement to the completely finished works.

Kant had another habit. In some of his lectures he used compendia: in logic, Meier's compendium, in metaphysics, Baumgarten's, etc. As he prepared his lectures he wrote in the margins, on empty pages, wherever there was room, everything that came to mind. Thus the textbooks he used over the years came to be covered with thousands of handwritten remarks. The notes from Meier's *Logic* alone comprise almost the whole sixteenth tome of the Academy edition of Kant's works. All of the notes fill ten volumes — that is to say more than all his published works. The notes are not dated; working from internal evidence, however, the editor of Kant's manuscripts, E. Adickes, managed not only to sort the material out thematically but also give it an approximate chronological order. The modern reader thus has a kind of scientific diary of Immanuel Kant at his disposal.

Here we have fragments which Kant was obviously planning to include in a new edition of *On the Feeling*. . . Kant thinks about freedom. He had considered freedom abstractly in his dissertation. Now, under the influence of Rousseau, the problem acquires a social dimension. Freedom is the antipode of slavery and dependence. Man is dependent upon many external things, but a more fierce and unnatural dependence than the dependence upon outward necessity is the subjugation to the will of another man. If I am free to begin with, nothing can reduce me more than the thought that in the future I will not be dependent upon my will but upon a caprice alien to me. Today it is bitterly cold; I can go out or stay put as I wish; the will of another determines not what is best for me in a given case but what seems necessary to him. I want to sleep, and he wakes me up. I want to rest, or to play, but he orders me to work. And if he is today inclined to be compassionate, who can guarantee that tomorrow he will not change his mind? When man is dependent upon another he is man no longer, he is nothing but another's possession. Slavery is the greatest evil of human nature.

"L'homme est né libre, et par-tout il est dans les fers. Tel se croit le maître des autres, qui ne laisse pas d'être plus esclave qu'eux" ("Man is born free and everywhere he is in chains. Whoever believes himself master of others will not cease to be more a slave than they.") — thus begins the *Social Contract*. Kant goes on to vary and deepen the theme set by Rousseau. It seems that freedom should elevate man above animals, but in truth man is lower than the lowest animal: it is easier to force man into subjection. In addition to enslavement by force there is yet another kind of slavery, more dangerous: it is slavery through blindness. It is grounded on the dependence upon things (for instance, comfort or pleasure), or upon ideas. Man can control things more readily than opinions, therefore the second form of slavery is the uglier and the more despicable.

Kant then follows Rousseau to the problem of alienation. He is not familiar with the term, but he grasps the issue correctly. What is at stake is the fact that antagonistic social relations convert the results of man's activity into something alien, inimical. Using the example of science, Kant shows how something good can transform itself into something evil.

"The harm that science causes man consists mainly in this: that the great majority of those who wish to make their mark in it do not strive to perfect reason but to pervert it, not to mention the fact that to them science is only an instrument whereby they satisfy their vanity . . . The learned men think that everything exists because of them. The nobility think so too . . ."[47]

Kant contends that the science of his time is suffering from two afflictions. One is narrow-mindedness, one-sided thinking; the other is the lack of a worthy goal. Kant will return to this theme time and time again. Here, for instance, is an eloquent fragment from another notebook: "Barbarous learning can be very assiduous, but it has no purpose, it is not dedicated to the best interests of the human species." Science needs a "philosophical overview." If the learned man has no "philosophical eye," he is a species of one-eyed monster.[48] When man shuts himself up within the prejudices of a branch of science he is in a dangerous state of deformity. "I call a learned man of that kind — a cyclops. He is a scientific egoist; he needs yet another eye to look at things from the point of view of other men. It is on this that the human character of science is grounded, i.e. judgments become permeated with the spirit of man . . . The second eye is thus the self-knowledge of reason, without it we cannot measure the greatness of our knowledge."[49] Kant set himself the task of overcoming the vices of contemporary science. "If there is one science man cannot do without, then it is the one which teaches him how to take his place in creation — from this one can learn what man must be in order to be a human being."[50]

This confession is of great significance for Kant. He breaks once and for all with the learned arrogance of the men of the enlightenment who are in love with their encyclopaedic knowledge which convinces them of the divine power of science. The value of science is in its ethical orientation; Kant wishes in the future to dedicate himself to the science of man. From now on man is at the centre of Kant's philosophical interests. Everything culminates in one question: what does man really need, how can one help him?

Would it not perhaps be best to abandon civilization? Everyone echoes Rousseau's statement. Fashionably dressed, perfumed, powdered ladies and gentlemen sigh about the pleasures of life in nature, free of their troublesome comforts.Could it be that the salvation lies in forgetting all the learned wisdom and seeking the truth in the Scriptures?

Then an experiment, quite unplanned, took place before their eyes. In January 1764 a bit of news alarmed the local intelligentsia: in a forest near Königsberg a "natural man" had turned up who had cast down the cloak of civilisation and returned to the sources of belief. It was a certain Jan Pawlikowicz Zdomozyrskich Komarnicki, fifty years of age, accompanied by an eight-year-old. Both were dressed in animal skins and went barefoot in all kinds of weather and in all seasons; they moved from place to place and lived from what they could get from their herd: fourteen cows, twenty sheep, forty-six goats. In the hands of the "goat prophet" was always the Bible, from which he quoted ceaselessly and without discrimination.

A pilgrimage to the forest began. Kant went there too. Certain circumstances became immediately apparent: the cause of it all was a serious stomach ailment Komarnicki had suffered seven years earlier—he had eaten nothing for twenty days whereupon Christ had appeared to him and led him to the path of truth. Now he lived on cows' and goats' milk; occasionally, on very holy days, he partook of meat. The police had sent the "goat prophet" back to Poland, whence he had come. The case was described in *The Königsberg Learned and Political Gazette.* Kant's *Enquiry into Illnesses of the Head* appeared in the same paper shortly thereafter.

The article was written very much in the light essayistic manner of *On the Feeling* . . . Kant's conclusion: the root of mental illness is the digestive tract. In his natural setting man was not disposed towards mental disorders as much as he is in the conditions of modern civilisation. Kant sees in the bourgeois social order perhaps not the causes but certainly the precipitating factors of mental illness, its aggravation and its continued duration. Kant's notion, expressed in modern terms, runs as follows: psychosis is a perverted protest against the perverted forms of social relations.

In 1764 Kant was forty years old. He was known, respected and admired. His lectures were successful, the auditorium was always full; he had passed several of his courses on to his students. His books were selling well and *On the Feeling of the Beautiful and the Sublime* had brought him the renown of a fashionable author.

But he was still only a *Privatdozent*; the university did not pay him a penny. In Berlin they understood the injustice of his situation. In August 1764 an enquiry came from the ministry of justice, which also dealt with educational matters: would *Magister* Kant consider taking the professorship of poetics and rhetoric which had fallen vacant two years before upon the death of Bock (the very same man whom Empress Elisabeth had rewarded financially and had had elected to the Russian Academy of Science)? In those days narrow specialisation had not yet become a cult. Theologians taught medicine, lawyers competed for positions in the natural sciences. Professor Buck, who had earlier been preferred to Kant, was a mathematician. Why should the philosopher Kant not try his hand at poetry; his interests were many and varied, and the work about the beautiful and the sublime proved that he was a man of taste. It certainly should not be difficult for him to edit others' verse or to compose a few of his own for official occasions: that was part of the duties of a professor of poetics.

Kant, however, declined. He had an aim and wanted to reach it by the most direct road: it could be long but it had to be straight. It was not one of his principles to make detours. He had been waiting for eight years, he could wait longer if need be.

The rejection earned him respect. In a special *Reskript* it was promised in the name of the king that "*Magister* Kant, the very learned and universally esteemed *dozent* at the said Academy, would be promoted at the first opportunity."[51]

A year later Kant would remember the promise. His financial affairs left something to be desired. He never borrowed money and was very proud of that; he used to say that he could say "come in" without fear whenever he heard a knock at the door knowing that it was not a creditor; but it was known in the city that *Magister* Kant occasionally had to sell his books. In October 1765 Kant found out that an assistant librarian's position was vacant in the library of the royal castle. There were several applicants for the post, but Kant reminded the authorities of the promise they had given him. In February 1766 he assumed the new post, without however relinquishing his university career.

The library occupied the lower floor of the castle keep (above it was the fire watch gallery). The repository was considered one of the best in the city: it contained sixteen thousand volumes. Our old acquaintance Bolotov visited the castle library frequently in his official capacity and left us the following description:

> The books are for the most part old, many are manuscripts; I saw very rare books penned by clerics of yore in very clear, orderly hand, illustrated with various figures done in brilliant colours; I found it astounding that many were attached to the shelves by lengths of chain: everybody can take them down from the shelf, leaf through them and read them, but they cannot be taken home. In summer the library is open throughout the week,

> and anyone can come in and sit there the whole day with his favourite book; they are only very careful to see that no book is taken out of the library. To make reading easier there are many long tables in the room, surrounded by benches; many, especially students and scholars, avail themselves of this opportunity, and I have seen ten to twenty men there, deeply engrossed in their books.[52]

Work in the library did not consume much time: it was now open only on Wednesdays and Saturdays from one until four in the afternoon. But the salary of the librarian was commensurate to the work—sixty-two talers a year. For a while Kant supervised the mineralogical collection of the library.

In the meantime his reputation was spreading. Lambert and Mendelssohn both approached him with the suggestion that they enter into a scientific correspondence. Johann Heinrich Lambert was a mathematician, astronomer, philosopher, four years Kant's junior and already a professor and a member of the Berlin Academy of Sciences. His letter was delivered by a theological student who had come to Königsberg. (Postage was high at the time, therefore letters were often entrusted to the acquaintances who happened to be going in the right direction.) Both of them were "doing research in the same field"[53] and yet they did not know each other, Lambert writes. He especially had their identical views about the creation of the Milky Way in mind. In 1761 Lambert repeated what Kant had said five years before. But, Lambert assured him, he had already thought of considering the Milky Way as a constellation in 1749, when, after a late supper he had beheld the night sky; he had then made a note of it on a piece of paper. (The information did not impress Kant; he held as before that the discovery was his and on every occasion claimed priority.) Besides, Lambert thought, they were even: Kant's recent thoughts about the philosophical method were merely an extension of what he, Lambert, had already published. In order to avoid such duplication in the future, Lambert suggests that they share their ideas before publication. The publisher Kanter had announced the forthcoming publication of Kant's work on the problem of method in metaphysics and Lambert could hardly wait to read it.

Kant was flattered. In his answer he called Lambert "the first genius of Germany." It was quite true that their methodologies were very close. (This is an important concession: Lambert considered himself an empiricist, he had entitled his major philosophical work after Bacon's book, *The New Organon.*) After many "twists and turns" he had proved the validity of his own method, in which observation occupied the central place. What Kant further wrote meant yet a more decisive "twist." Philosophy was in a predicament: it did not have a universally binding set of rules whereby its propositions could be put to test. As for his work on the philosophical method there was obviously a misunderstanding: the publisher was making publicity too early. A work like that, which could indeed become the major goal of his future endeavors, was still far from completion. He was, however, sending Lambert several minor works on theoretical and practical metaphysics. This was the first mention of Kant's "major work." He had in mind the *Critique of Pure Reason.*

Lambert's next letter was a tractatus in itself; in thirteen points he laid out his understanding of the philosophical method, then in five points clarified the reciprocal relationship between matter and form, and completed the points with nine theses. Lambert was waiting impatiently to hear the opinion of his Königsberg colleague.

In the seventeenth century there were no scientific presses; scholars exchanged their ideas in correspondence. Spinoza's correspondence is a unique philosophical journal. Kant's letters are different in that respect: with a few exceptions they were not meant for publication — they were lapidary, somewhat detailed, dry reflections of his scientific and private preoccupations. In the eighteenth century the epistolary genre became one of the favourite forms of literary activity. Kant was not susceptible to its enticements; in correspondence he did not exert his talent and did not distinguish himself by unusual assiduity. He wrote to Lambert again only four and a half years later.

Nevertheless the amicable relationship did not suffer a setback. When his student, Jensch, travelled to Berlin Kant asked him to convey his friendliest greetings to Lambert.

Lambert received another answer from Kant: via Mendelssohn he received Kant's new work *Dreams of a Visionary Explained by Dreams of Metaphysics.* That was an essay rather than a treatise, dedicated to the activity of an unusual man. Emanuel Swedenborg. The Swedish philosopher and mathematician gained an early fame through his work in mechanics, mineralogy and mining and was elected to the Petersburg Academy of Sciences. In his old age Swedenborg declared himself a visionary whom God himself ordered to found a new church. He was convinced that he stood in a close relationship with the souls of the dead, i.e., that he received guidance from them from the other world and gave them the news of this world.

Kant had first heard about Swedenborg from an old student, a Danish officer. The philosopher began to gather details and compare evidence. He wrote to Swedenborg, but received no answer: everybody was going to receive the answer in the form of a book. Kant asked an acquaintance to call on Swedenborg in Stockholm. The man obliged and heard from the mouth of the visionary the following: God had given him the ability to traffic with the souls of the dead at will. Proofs? Everybody knew them.

A silversmith asked the widow of the Dutch Ambassador to Stockholm to pay for a silver service her late husband had ordered. The widow knew that her late husband had been punctilious in meeting his obligations; she was certain that the bill had been paid in full but could not find the receipt. Deeply disturbed, for the sum had not been paltry, she invited Swedenborg to see her. If, as everybody said, he was truly given the unusual gift of communing with the souls of the dead, would he kindly ask her husband where the receipt was to be found? Swedenborg found it easy to accede to her request. Three days later a small group gathered at the residence of the lady for coffee. Swedenborg came as well and calmly told the woman that he had spoken with her late husband. The bill had been paid seven months before his death, and the receipt was in the cupboard of a room on the top floor. The lady said that the cupboard was quite empty. Swedenborg answered that — according to the words of the late husband — the left drawer had to be taken out, its bottom removed, and there, in the secret compartment, was some confidential Dutch correspondence and the receipt. Everybody went upstairs, the cupboard was opened, the directions were followed and the secret compartment of which the mistress of the house was ignorant, was discovered. In it: the papers in question.

Another instance appeared to Kant even more convincing, more free of doubt. It happened at the end of September 1756. Swedenborg was the guest of a merchant, some fifty miles from Stockholm. Suddenly he turned pale and said that fire had broken out in Stockholm. A little later he said that his friend's house had burned down and that fire was now approaching his own house. At eight o'clock he exclaimed joyfully: "God be praised, the fire has been put out near my house!" Two days later a courier arrived from Stockholm with an account of the conflagration which bore Swedenborg's vision out down to the last detail.

There was a time when Kant did not doubt the veracity of this or of the other account. In a letter to Miss Knobloch[54] who had asked him for an explanation, Kant repeated these stories as fully believable, even though they puzzled him. The date of the letter is illegible: in the Academy edition it is dated 1763. That cannot be correct: already a year before that Kant doubted resolutely the existence of spirits (we can recall Herder's notes of Kant's lectures). The letter to Miss Knobloch reflects an earler Kant. It seems that Borowski was right: he was the first to publish the letter and he dated it 1758.[55]

Kant's early belief in spirits is not surprising. The eighteenth century swarmed with seers, "magnetisers," alchemists. Some of them really were masters of the art of hypnosis, others were simply charlatans. We shall speak about Cagliostro later on. In the years of the Seven Years War his predecessor — Count Saint-Germain — appeared who claimed to have lived for an extraordinarily long time: he was present when Alexander the Great entered the city of Babylon and he had known Jesus Christ personally. Aside from maintaining contacts in the spiritual world the count liked political intrigues; the latter love was the grave of his reputation. He strove to use his influence at the French court to benefit Prussia. Suspicions about his secret connexions with Friedrich II grew, and when it became clear that he was going to be arrested the adventurer fled to Germany.

Kant was quickly convinced of the fact that Swedenborg was a fraud; he ordered Swedenborg's *Arcana Coelestia* (*Heavenly Secrets*) from London and paid a pretty penny for it. His interest was so great that he was willing to pay the high price from his meager income. Kant finally received eight volumes "full of nonsense"; he expressed himself similarly in *Dreams of a Spirit-Seer*, where he subjected Swedenborg to a sustained analysis.

Kant tells the two stories there we already know, only now his tone is ironic. "There is no doubt that someone will ask me why I have spent time with such worthless drivel, why I am peddling stories which a reasonable man will refuse to hear to the end, and, ultimately, why I have made them into the text of philosophical investigations,"[56] Kant asks the rhetorical question. And then he provides a melancholy answer: "The line between foolishness and reason is so faint that if one travels in one of them for a long time it is difficult not to stray now and then into the other." But there is a line between them, and the reader should not be judged harshly if he sends the visionaries to the physician instead of considering them semi-denizens of the other world.

Swedenborg and his following are not all that is at issue. Kant places adept speculators in metaphysics on the same level with visionaries. They are visionaries of reason, just as the latter are visionaries of feeling. Metaphysicians dream, too: they take their ideas to be the primordial order of things. The philosopher does not begrudge them their "discoveries," he is only afraid that a reasonable but not very polite man might tell them what Tycho Brahe's coachman told the astronomer who thought that it was necessary to drive at night and by the light of stars; "My dear sir, you may know the heavens well, but here, on earth, you are a fool!"

This is, as it were, Kant's last word on Wolffian metaphysics. He is amused not only by visions but also by speculative prestodigitations; he asks that science be based on experience and on experience only, the alpha and omega of knowledge.

Kant says farewell to metaphysics, but he cannot part with it. His confession: it was his destiny to fall in love with metaphysics "although she rarely shows her gracious visage to me."[57] This unrequited love lasted many years. Throughout his teaching career Kant taught a course in metaphysics ("according to Baumgarten"), he was troubled by those "damned" metaphysical questions — about the nature of the world, God, soul. But the longer he worried, the clearer it became to him that answers could not be obtained by speculative approaches. Therefore Kant dreamed about the re-education of his love: he wished to see her as a companion of wisdom through whom he would come to know the naturally established limits of human reason.

In the preface to *Dreams of a Spirit-Seer* there is a third object of philosophical irony: the church. "The empire of shadows is a visionary's paradise . . . Only holy Rome disposes of the wealth of her provinces; the two crowns of the invisible empire support the third — the decrepit diadem of earthly greatness, and the keys that open the gates of the other world open with equal ease the coffers of this one."[58] Kant speaks of the Catholic church, but that does not mean that he speaks in the name of Protestant orthodoxy. With this too, he will later have many difficulties.

In 1766 Kant changed his quarters. The old flat, at Kneiphof, had been near the university, in the Magister's Mews. Its windows opened on the river Pregel, and the noise of passing barges disturbed Kant's concentration. He now rented the left half of the second floor in the house of Kanter, the bookseller, at Löbnicht. Kanter was an educated and enterprising man; he did not only sell books, but established a publishing house and library as well, organized a lottery, bought a newspaper and founded a paper factory. He spent half of his time on the road, enlarging the sphere of his activity. It was from him that Kant acquired Rousseau's *Social Contract* and, it seems, *Emile*.

Kanter's bookshop was a local curiosity: it was something between a business and a literary club. Here prominent persons met, read papers, learned of new books, conversed, and sometimes waited on customers. Twice a week students were allowed to read here free of charge. Hamann was also a regular in the bookshop. *Magister* Kant could claim the privileges of a friend and lodger and take any book he wanted. From the fall of 1768 on, his portrait (a work by J.G. Becker) graced Kanter's office together with portraits of other renowned Prussians.

It was not only in Prussia that Kant was well-known. In 1769 Professor K.R. Hausen of Halle intended to publish *Biographies of Renowned Philosophers and Historians of Germany and the World in the Eighteenth Century*. Kant was also included in this collection, and the author approached him with a request for the necessary material.

At about the same time Kant received the offer of a chair in Erlangen, in the margravate of Ansbach. The Chairman of the board of trustees was enthusiastic about *On the Feeling of the Beautiful and the Sublime* and invited Kant to accept the newly founded chair of theoretical philosophy. To have such a chair was one of Kant's favourite dreams. The salary was five hundred guldens and all the wood he would need for the winter. The temptation was great, and Kant gave something in the nature of a tentative acceptance. Immediately the letter of appointment was produced, the flat was already waiting (for the first time four rooms); students of the university were jubilant and busily readying everything for a festive reception. Kant remembered another reception and promptly declined.

The second offer of a chair (in January 1770) came from Jena. There the vacancy was less attractive — the second chair in philosophy at a salary of two hundred talers; Kant was guaranteed, however, that he could easily earn another one hundred and fifty talers by tutoring and that prices in the duchy of Weimar were lower than anywhere else. A much weightier argument would have been their assurance that Weimar was soon to become the centre of literary Germany, that in it the great friendship between Goethe and Schiller would be born. Of course, no one knew that then; Goethe had just begun to write, and Schiller was still a schoolboy of eleven.

In his rejection of the offer from Erlangen, Kant pleaded his ties to the native city, his frail health, and a possible vacancy at the University of Königsberg. The last was not only wishful thinking, it was soon to become reality: the seventy-nine-year-old professor of mathematics Langhansen had been bed-ridden for a long time. He died in the middle of March. Kant honoured his passing with a mournful stanza but the following day wrote to Berlin reminding the authorities of his existence. He would soon be forty-seven and he was not yet a full member of the faculty. The vacancy, it was true, was not in his area of specialization but that problem could be remedied by a simple exchange of positions. Christiani, the best mathematician at the university, taught moral philosophy; logic and metaphysics were being taught by Buck, the mathematician. Both would be happy to take the vacant chair, and he, Kant, could then take up teaching what he was most suited to teach.

Two weeks later he had the answer: by a royal order of 31 March, 1770, Kant was appointed full professor of logic and metaphysics. (The salary was lower than the salary offered by Jena — 166 talers and 60 groschen — so Professor Kant retained his position at the library for another two years.) But he had reached his goal. On 2 May the senate of the university installed him in his office. There was still another formality to be completed: he had to defend yet another dissertation. He had defended a dissertation (*Habilitationsschrift*) fourteen years earlier (*Monadologia physica*), but the time which had passed was too long, and the rules on the matter demanded a new dissertation.

Kant spent the whole summer working on this dissertation. The debate took place on 21 August: it was the fourth dissertation that he defended. His opponents were two candidates and a student; the respondent was Marcus Herz, his friend and student. The young Lenz, the future poet of "Storm and Stress," attended the dispute and celebrated in verse the important event in the life of his teacher.[59]

. . .
Honour crowns the head of man
Whom virtue, wisdom hold in span,
Who teaches man by word and deed,
Whose life's a lesson of his creed.

His righteous eye to falsehood blind,
True form of wisdom knows his mind,
So oft the dreaded mask of lies
He stripped away for our eyes.

. . .

Ever we shall him extol,
Whose wisdom nurtures our soul,
And as he taught us we shall live,
The gift he gave us we shall give.

Ye sons of France, look not with scorn
At North, where genius was born,
Never doubt, while Kant yet lives
His life your queries answer gives.

The dissertation was entitled *De mundi sensibilis atque intelligibilis forma et principiis (On the Form and Principle of the Sensible and Intelligible Worlds)*. In it, Kant had a new perspective, the result of the work done in the year before. Empiricism, which inexorably lead into scepticism, had been replaced by a peculiar dualism. Kant was no longer interested in the relationship between sense data and intellect — he had already considered the two aspects of mental activity in a variety of ways. "The sources of all our representations" — so Kant wrote in a note from the crucial year, 1769 — "are either our senses, or our understanding and our reason. The former give us the material of phenomenal knowledge which express the relationship between the object and the unique characteristics of the apprehending subject . . . The latter are related to the object itself."[60] Senses had to do with appearances, phenomena; the intelligible object, which can only be grasped by reason, Kant called noumena.

The world considered as a phenomenon exists in space and time. But neither of the two (space and time) exists in itself (*an sich*); they are only the subjective conditions present in understanding (*Verstand*) from the beginning so that it may coordinate the sensibly apprehended objects. In the noumenal world, the sphere of the things in themselves (*an sich*), there is no time and space; it is therefore unacceptable to ask where God can be found and why he did not create the world eons earlier.

In 1769, the year of Kant's great change, L. Euler's book *Letters to a German Princess* appeared; the writer answered the question of the relationship between the body and soul by stating that the relationship could be thought, but not apprehended sensibly. Kant took up the idea of the great mathematician; it seemed to him that it offered the solution to a problem he had been examining for a long time. Kant turned the answer on its head: there were objects man could intuit, but could not think. Spatial relationships were in that category.

Until recently Kant had been asking that metaphysics rely exclusively on experience, now he was interested in preventing it from becoming too dependent on experience: the principles of sensual knowledge should not go beyond their own province and become contiguous to the sphere of understanding. In a letter to Lambert, which he had finally written and sent to Berlin together with a copy of his dissertation, Kant proposed that a new discipline be established — a "general phenomenology" with the task of delineating the limits of sensual knowledge so that that knowledge would not be applied to the objects of "pure reason." (The reader should note that the term "pure reason" makes its appearance here for the first time!)

Copies of the dissertation were also sent to Mendelssohn and the celebrated aesthetician I.G. Sulzer. Three letters in quick succession came from Berlin. Lambert was the first to answer. His long epistle began with the lament that science everywhere was in decline, everybody was interested in the arts and aesthetics, students were reading only novels and poetry. When they left the halls of the university and went into the world they had to be reeducated, or simply begin anew. He, Lambert, was working on the establishment of a press where the spirit of strict science would rule; works of heretical and capricious content would be barred entrance. He found Kant's dissertation quite splendid, but did have a few questions and remarks. Kant sets sensual knowledge and the knowledge of understanding poles apart; do they never meet? Then, there was the problem of time. It underlies precisely those changes which occur in objects. If the changes are real, then time is real as well. The same applies to space. In general these are difficult matters, but man must not strive for absolute clarity; there must always remain areas of indeterminacy. Appearance, if constant, must be taken as true. "The language of appearances" can serve us just as well as "the language of truth."

Then came the letter from Sulzer. The master of aesthetics informed Kant that he was busy with the publication of his new treatise on the fine arts; that did not prevent him from giving careful thought to all the "new concepts" he found in Kant's work. These concepts appeared to him essential and important. There was only one small point on which he did not agree — the problem of space and time. (On this issue Sulzer agrees with Leibniz and Wolff, who grant primacy to duration and extension.) He waited with great impatience for Kant's work on the metaphysics of morals. He was interested in the physical and psychological differences between a virtuous soul, and a vicious one. He himself had given some thought to the problem, so important for education, but at present he had so little time. He sincerely wished Kant would be able to continue the work so splendidly begun, he wished him health and the strength to complete the work successfully.

At the end of 1770 Mendelssohn's letter arrived. He had read the dissertation with great pleasure, although of late his nerves had given him trouble and he found it difficult to concentrate. To benefit metaphysics, which was in dire straits, it was highly desirable for Kant to share his thinking with others: life was so short and the end was so quickly upon us without our having succeeded in doing everything to the best of our ability. Do not fear repeating what has already been said; in relation to your own ideas the old assumes a new appearance, opens new perspectives, "because you have a great talent to write for a broad public."[61] Mendelssohn made a few remarks on the text; like the other two before him he could not agree with Kant's treatment of space and time.

Kant did not answer any of these letters. Each letter had set before him new problems and forced him to think more, for a time without result. Half a year later he wrote about it to his friend Marcus Herz in Berlin, begging him to convey his apologies to his correspondents with the explanation that he was not well, and that he was working on another book, *The Limits of Sense and Reason*.

Eight months later Kant wrote to Herz again and begs him again to offer his apologies to Lambert, Mendelssohn, and Sulzer. The letter contains more than the bland statement of fact that he needs more time to consider the objections presented and to think about his own ideas. The date of this letter (21 February, 1772) must be considered the birthday (perhaps, more accurately, the day of conception) of Kant's major philosophical work. The day of delivery is still far in the future, but the embryo has formed and begun to develop. And although Kant calls his future book as before, *The Limits of Sense and Reason*, the expression "the critique of pure reason" appears somewhere in the text. The future book is to provide "the key to the whole secret of metaphysics." No one up to this point has answered the question of how concepts come to be. It is easy to see how matters stand with sensible representations: it is a passive impression of the object. That cannot be said of concepts. Here we have to deal with the products of the intellect (*Intellekt*) which at the same time correspond to the inner order of things. How is this possible?

The answer will be provided in the book on which he is now working, Kant says. The first part will be completed within three months, he assures Herz.

Almost two years pass. And Kant has written another letter to Herz: "You are diligently but vainly searching the catalogues of published works for a certain name under the letter K. After all the work and worry it would be very easy for me to publish something that is almost complete. But as I have gone so far in my attempts to reshape science over which a good half of the philosophical world has laboured for so long that I can now see within my reach an outline which will fully reveal the as yet opaque enigma and bring thus isolated reason (*Vernunft*) within the compass of firm and easy rules. I shall remain steady in my intention not to yield to the urge for a fast and easy fame; I shall stand my ground until I have completed the whole work. . . . I hope that I shall be able to send the finished work off by Easter."[62]

Thus Kant at the end of 1773. Three years later he writes to Herz: "I do not think I shall be able to finish the work by Easter; I shall have to use part of the coming summer."[63] Another two years later Kant was hopeful that he would finish the work "by summer."[64]

The book did not go well. Never before had Kant worked so slowly. He kept missing a major issue. The truth he woud bring to the light would turn into another insoluble puzzle. And, he was not only working on the *Critique of Pure Reason*. Kant did not devote himself heart and soul to gnoseological problems after 1770 and he did not use all his strength in fruitless attempts to solve them. As before, lectures took up much time and energy. It is true that his teaching load was lightened, on an average fourteen hours a week. But he had introduced new courses: mineralogy, anthropology (soon to become one of his favourite subjects), and pedagogy.

Kant's pedagogical views were formed under the influence of Rousseau and by his own attempts to realize Rousseau's ideas on the German soil. In 1774, with the help of the local ruler, the celebrated pedagogue Johann Bernhard Basedow opened an educational institution — the *Philanthropin* — in Dessau. The institution was going to reform all of pedagogy. Education in the *Philanthropin* was decidedly cosmopolitan in character: instead of Christian doctrine children were taught a general, "natural" religion, based on virtue. Natural and exact sciences were paid a great deal of attention. Children learned in play, without drilling and without punishment. Free time was not supervised; educational achievements were rewarded with medals. Much time was also spent on the cultivation of the body. Admission to the institution was limited, and the education there very expensive. Poor children were taken on only as servants.

Kant was very much attracted by the new movements in pedagogy. Although he had limited his literary activity, the philosopher published two notices in favour of the *Philanthropin* in the local newspaper. Kant called not for reform, but for a revolution in the whole system of education. On Kant's urging, his friend Motherby sent his six-year-old son to be educated in Dessau. That gave Kant the opportunity to enter into correspondence with the directors of the institution — Basedow and Campe. Several years before Wieland had asked Kant to help him circulate his newly founded paper *The German Mercury* in East Prussia. The philosopher had then declined, suggesting that Wieland get in touch with the publisher Kanter instead. But with the journal that the *Philanthropin* was going to publish it was a different story; Kant himself took up the matter. The result was modest: altogether ten Königsbergers and fifteen Lithuanians subscribed to it. It was not different in the rest of Germany: the new pedagogical ideas were not very popular.

In 1774, when the *Philanthropin* was established, pedagogy began to be offered at the university for the first time. Seven professors of the philosophical faculty taught the subject, taking turns. Kant's turn came in the winter of 1776 (and then again six semesters later). Kant used Basedow's book as his text, entering corrections and addenda in it, as was his habit. Out of that grew an independent work, *On Pedagogy*, which Kant's student Rink published shortly before the philosopher's death.

Two human inventions can be considered the most difficult, Kant writes there: they are the art of government and the art of education. But it is precisely upon them that society rests. "Man can only become man by education. He is what education makes of him, nothing more."[65] Therefore a philosopher who is concerned about the fate of man must face pedagogical questions.

Parents and rulers corrupt children. Parents only worry about their offspring being able to make it in life, rulers only consider their subjects as instruments of their sovereign power. Both are products of the education of an existing world order. But genuine education must prepare man for the future, for a drive to improve the condition of the whole humanity. Only then will he be able to make it come about. How will he go about doing it? The ruler is always alone, and a lonely tree grows crooked. Basedow's hopes for the monarch are in vain. The school system should not be in the hands of the state. In Kant's view, every culture begins with the initiative of individuals.

Man can be either drilled or cultivated. The main aim of education must be to teach thinking. Man must consciously pass through four stages of education: he must learn discipline, he must learn certain skills, he must learn how to conduct himself in society, and he must become morally good. The last stage is the most difficult. We live in an epoch of discipline, culture, and civilization — but we are far from being morally good creatures.

One of the most important tasks of education is the joining of the compulsion of laws to the ability to exercise freedom. The child must be given his freedom from the earliest moment possible (except in those cases in which he is likely to harm himself or others). He must be shown that he can achieve his aims only if he allows others the possibility of achieving theirs. (Pleasure only as the reward for obedience!) Public education's superiority consists in its allowing the child to learn to adapt his rights to the rights of others. In this way the child attains a notion of duty. The personality, that is the freely acting individual, must be educated in the spirit of its own human dignity, and in the consciousness of the resulting duties of a member of society.

Although Kant had no children, he did have his opinions on how they should be raised, beginning with infancy. The philosopher advises mothers not to use nursemaids, but to nurse their infants themselves. The infant should not be swaddled, which only produces fear and insecurity. Try diapering a grown man, he says, and he too will cry like an infant. Children should not be rocked. Try it on yourself, he suggests: the back and forth motion will make you sick. Children are rocked so that they will not cry but crying is healthy: it develops the lungs. The first spoiling of the child's character begins when his parents rush heedlessly to soothe him the moment he begins to cry. If one pays no attention to his crying the child will stop it of his own accord: no one is disposed to expend energy in vain. When parents teach the infant how to walk they should not use a harness, which compresses and deforms the chest. The infant should in general be left to his own devices.

All this should be done in such a way that discipline does not suffer. The will of the child should not be broken; it should be molded so that it learns to yield to natural obstacles. Children should be taught to work as soon as possible; nothing is more harmful than sloth. An unspoilt child is eager to do things which use his powers and his will. In the formation of eating habits the child should not be permitted to choose his own food; that may make him greedy. If there is no strong reason to deny a request it should be granted; if there is a strong reason then the parent must not permit himself to be cajoled into acceding. A no must remain a no, once and for all.

Following the example of the *Philanthropin* Kant strongly recommends physical exercise, which was a pedagogical novelty in his day. Running, jumping, weight lifting, swimming, marksmanship — all these benefit children. Games in the fresh air are also beneficial. A child must know how to play. But play must not be confused with work. Kant does not subscribe to Basedow's theory that children learn while playing and cavorting about. The most important task of the school is work, it must be made to lead to a determined goal. Play has its end in itself, its most important ingredient is progress, it has no other aim. Grown-ups should know the difference between work and play: they no longer ride hobby horses but horses. It is extremely harmful to teach the child to regard everything as a game. School is an education in the power of compulsion, although its aim must be to lead man to freedom.

According to a Latin proverb we learn only as much as we can retain in memory. Therefore it is extremely important to train the memory. Systematic reading and repetition of what is learned and the learning of foreign languages strengthen the memory; reading of novels, on the other hand, weakens it according to Kant — therefore they should be taken out of the hands of children. When they read a novel they build a new one out of it because they misunderstand it; then they sit lost in thought and daydreams. Absentimindedness is the enemy of all education.

The educator must not abuse punishment. It must remain an extreme measure, to be used very carefully so that children see that in the last analysis it is applied for their good. Frequent punishment accomplishes nothing; the same holds true with the punishment applied in a fit of anger.

It is most difficult to develop a moral culture in man based on principles. Discipline alone avails here not at all. These principles are laws, subjective, it is true, springing forth from the character, convictions and reason of man. To develop character it is necessary to cultivate three essential features: obedience, truthfulness, and sociability.

Kant speaks finally of sexual education. The age of thirteen or fourteen is a crucial one in a boy's life: sexual urges develop. They develop inevitably, although as yet without an object. For this reason one cannot preserve the youth in ignorance and innocence. Silence only magnifies evil; we see that from the education of previous generations. It is therefore proper to speak with the youth about everything, clearly and openly. Naturally, the subject is delicate and cannot be the subject matter of a general conversation, but it is always possible to find an appropriate form of discourse. In keeping with the views of his time Kant suggests that the pedagogue develop the fear and repugnance of masturbation in the youth. "Physical effects are generally harmful, but from the moral standpoint they are even worse. Man here tramples the laws of nature underfoot; the rage to satisfy the inclination cannot be stopped because it offers no genuine release. Teachers of growing youths have asked the question: should the youth be permitted to enter into a relationship with the opposite sex? If one must choose, then that is by all means to be preferred. The former is unnatural, the latter is not."[66] Unfortunately it is easier to procreate than it is to bring up and educate children. Consequently the youth's first responsibility is to wait until he can marry. A happy marriage is the reward of immaculate conduct.

How did Kant's ideas about the education of youth fare in his own pedagogical praxis? Kant dealt with students, with the young who had yet to pass the first hurdles of education. The duty of a university lecturer was to transmit knowledge and teach independent thinking.

The latter is, of course, more important than the former: knowledge can be obtained, but the free use of reason is difficult to acquire. Mechanical memorization and grind are harmful. Kant was unhappy when his students took down everything he said; that angered him. When he began a lecture he strove first of all to make himself understood. When the thread of understanding broke he stopped and went back to the beginning. He aimed his lectures at the average student. "I do not lecture to the genius, because he will cut his own path without my help; nor do I speak to the fool, because he is not worth it; I concentrate rather on the student who stands in the middle and desires to prepare himself for his future profession."[67]

He usually chose a listener who sat near him and read in his face whether he was being understood or not. Once he had begun to unfold his thought, the slightest disturbance in the auditorium could interrupt his train of thought. He was once distracted and lectured worse than usually; he then confessed that he had been distracted by a student who sat right in front of him with a button missing on his coat.

He lectured sitting behind a low lectern on which there was a page of notes or the textbook filled with observations and comments. He did not use the textbook to regurgitate or systematise material; he used it to incite controversy. Lectures then became dialogues. Difficult passages were elucidated with illustrations, pregnant remarks and jokes. Jachmann, who heard him in the seventies, tells that "his quips were light, droll, profound. They were shafts of lightning in the clear of the day, and he spiced not only his everyday conversation with them but also his lectures. His quips gave the serious, deep thought an attractive form and pulled his listeners down from the high spheres of speculation to the earthy regions of life."[68]

In an examination the student and the teacher changed places. Then it was the student's task to explain the heart of the matter. A smooth style and a command of the norms of exposition were of secondary importance. If a student found it difficult to express himself clearly, Kant did not think it said much of his intellectual powers. The famous scholar Clavius had been expelled from school because he had not been able to write imitative compositions. As the result he quite by accident began to take an interest in mathematics and discovered in it his great talent.

In philosophy mistakes are even more possible: stupidity can easily pass for wisdom, and a new thought can defend its claim with difficulty. That is why Kant liked to repeat that he did not teach philosophy but the art of correct thinking. To master philosophy is to master the subjective side of reason: it is to remember what someone thought at some time; to learn to think correctly is to give one's knowledge an objective character, to acquire the ability to judge opinions of others. Kant laid out his projected major work not as an enumeration of dogmas or a compendium of rules, but as a critical review of what had been done and what still needed doing.

Reason has always existed, just not always in the rational form.
— K. MARX

CHAPTER THREE

The Self-Critique of Reason

Years passed. Kant had not published anything in a long time: after the dissertation *De mundi sensibilis atque intelligibilis forma et principiis* and the two articles about the *Philanthropin*, there was only the review of Moscati's book *On the Difference in Bodily Structure between Man and Animal* and the announcement of his lectures for 1775 (*On Different Races of Man*). The silence lasted eleven years.

"Tell me in a few lines," Lavater pressed him, "have you forsaken the world? Why is it that so many who cannot write, do, and you, the master of the pen, do not? Why are you silent, why do you give no sound of yourself in these new times? Are you asleep? Kant, — no, I shall not praise you, — but tell me why are you silent? Or rather: tell me that you are about to speak."[69] When Lavater spoke of the "new times" he had in mind "Storm and Stress," a movement which had begun to spread throughout the intellectual life of Germany.

German literature had come to life. By the middle of the century there was already the rich promise of Gotthold Ephraim Lessing, who would shape the taste and the thinking of the coming age. He articulated a new dramaturgy and confidently raised his voice in literary criticism. His *Letters on the Most Recent Literature* came out. The seventeenth letter, dedicated to Shakespeare, is well known. The great dramatist had not been entirely forgotten in his native land but his fame had paled. On the continent he was absolutely unknown. Lessing thought him superior to Racine and Corneille; he thought that Shakespeare truly carried on the traditions of classical tragedy. Go and learn from Shakespeare, Lessing told the writers of his land. "Storm and Stress' could begin.

The movement was heterogeneous and full of contradictions: rebellion coexisted with political indifference and conservatism, love of nation with extreme individualism, religious scepticism with exalted piety. But all were united in an interest in man, in his unique spiritual world. Suddenly the secret depths of the individual were revealed. This discovery was deemed far more significant than the discovery of America. Everyone thought and spoke of one thing only; the liberation of the personality.

It is true that the "Storm" howled exclusively within the literary realm, and the "Stress" never broke out of books and journals. Among the leaders of the movement we find Kant's student, Herder, who was now a respected man of letters. He had left Königsberg for Riga, where he was a pastor and teacher in the parochial high school. On the publication of his first literary-critical works, which were to make him famous, Kant sent greetings to his pupil. He received an answer half a year later: it was courtly, warm, detailed. The pupil wrote that he had chosen a path different from the path followed by his teacher; he had preferred priesthood to teaching in the university because it seemed to him that as a priest he would be in direct contact with the simple men of the nation: after all, philosophy existed in its fullest sense for their sake. Herder did not stay in Riga long; he undertook a journey to Paris and upon his return to Germany he took the post of a councilor at the consistory at Bückeburg. There he spent the years 1771 to 1776 trying to find his way as a writer. In addition to his official duties Herder read philosophy of history, developed an enthusiasm for Shakespeare, introduced folk art and poetry to the educated world, and studied the Bible with much diligence. He published *The Oldest Document of the Human Kind*, where he considered the Scriptures as a divinely inspired work. Hamann, who shared Herder's religious enthusiasm, showed the book to Kant. The philosopher had little sympathy for Herder's attempt to represent the biblical text as the highest truth. His opinion fortunately was never expressed outside a private exchange of letters, and Herder never heard it. Besides Herder and Hamann, Lavater and the young philosopher Friedrich Jacobi also belonged to the religiously oriented branch of "Storm and Stress."

The other extraordinary head of the movement, Goethe, did not share the pious ambitions of his friend and teacher Herder. He blasphemed and wrote rebellious verse: "I know nothing more pitiful under the sun than you, oh gods! I should honour you? Why?" These words from Goethe's *Prometheus* mark the poem as a poetic manifesto of the "left" wing of "Storm and Stress." Among the radical "stormers" we find Kant's old student Lenz, and the future Kantians, Klinger and Bürger. Kant himself, who had been a forerunner of the movement, paid little attention to the "literary revolution" now that it was in full force. He was again a step ahead of everybody else, preparing a philosophical revolution.

On the other side of the Atlantic a political revolution was flaring up: the storm and stress that fell upon the British rule in America smashed a colonial empire for the first time in modern history. In July 1776 the American colonies of Britain declared themselves independent. The battles were fought under banners bearing the same mottoes held by German "stormers": freedom for the people, freedom for mankind. Kant read newspapers and his sympathies were with General Washington and his fellow combatants.

Once, in a conversation in the street (it must have been before the War of Independence) Kant expressed his opinion about the coming conflict in North America. One of the men present declared that as an Englishman he was offended and he challenged Kant to a duel. Now, the philosopher did carry a sword, but he did not know how to use it; his weapons were thought and word. He continued to speak, clarifying his position with so much conviction that the opponent in the end admitted that he had been in the wrong and offered his hand in peace. That is how Kant met Green.

Here is one of Kant's reflexions from those days: "In the contemporary history of England, her oppression of America is a great blow to the cosmopolitan view. The English wish the Americans to be the subjects of subjects and to bear the burdens of others. It is not good government, but the kind of government which is at issue."[70] The last sentence should be remembered: the thought would reappear time and again in Kant's future works and culminate in a comprehensive theory of the state. As yet it was only touched upon in passing; the philosopher was working hard on epistemology and finding the going extremely slow.

When one of Kant's pupils revealed to the professors at Göttingen that in the desk of his teacher there lay a work which would "rattle" the learned gentlemen to the core, they burst out laughing: one could hardly expect such a thing from this philosophical dilettante. (Among those present and laughing was also Professor Feder, of whom more later.)

Drafts accumulated in Kant's desk. There was no sign of the book's publication. He had stopped mentioning the book in his letters to Herz.

Kant was restless. In 1775 he changed lodgings again. The reason this time was that he was bothered by the neighbour's rooster, who crowed under his windows. The philosopher offered the neighbour money if only he would cut the rooster's throat, but the master would not hear of it. How can a bird bother any man, let alone one who is a famous sage to boot? The neighbour had no inclination to humour the philosopher. Thus Kant had to leave Kanter's house.

The move was not a great problem, for Kant's possessions were few: "Keys. A bureau. An inkwell. Pen and knife. Paper, manuscripts. Books. Slippers. Boots. Fur coat. Hat. Nightgowns. Napkins. Tablecloth. Towel. Plates. Keys. Knives and forks. Salt shaker. Decanter. Wine and beer glasses. Bottle of wine. Tobacco, pipes. Tea service, tea. Sugar. Brush."[71] This is the list of modest furnishings compiled by the philosopher on the eve of his move to his new lodgings. There, at Ochsenmarkt, he lived in quiet. — But still no news of the book.

During these years the head of the Prussian ministry of education was Baron von Zedlitz — a carbon copy of his monarch: a small enlightened despot. To his subordinates he emphasized that philosophy was worthy of respect; the student had to bear in mind, the minister said, that upon the completion of his studies he would only be a physician, a judge, a barrister, etc., a few hours each day: He was a man, however, the whole day through.

Therefore, in addition to specialised knowledge, the university had to provide the student with a solid philosophical orientation; this had to be genuine wisdom, not empty wordmanship and sophistry. The minister was a believer in Kant's philosophy.

What was the significance of an adherence to a certain system of thought for a man in power? In December 1775 a royal charter ordained the instructional order of the University of Königsberg. "Although we do not wish to regulate every educational endeavour, we deem it necessary to curb the dissemination of some which have proven to be altogether useless."[72] In the name of the king, von Zedlitz forbade the teaching of Crusianism, because it had become outmoded and superfluous; *Dozent* Weymann (Bolotov's beloved teacher and Kant's enemy) had to leave the university. In the royal charter Kant was given every sign of confidence and respect. The minister had his plans for the Königsberg philosopher.

It was 1777. One lovely summer morning an unknown man appeared in Kant's auditorium. He was too old to be a student and he did not look like an official; he was short and sported a luxuriant beard. At first he was regarded in silence, then jokes began to fly; soon there was noise, clamour. The unknown man remained unperturbed and quietly sat down on an empty chair. Kant appeared and began to lecture. During the pause the unknown man approached the lectern — the clamour rose again and was cut short: the professor took the unknown man in his arms. It was Moses Mendelssohn.

Kant was genuinely glad to have met a man with whom he could discuss philosophy on an even footing. He invited Mendelssohn to his house and begged to be forgiven the modesty of his table, but the host, like the guest, was interested only in talk. Mendelssohn had come to East Prussia on some personal business. He had, however, been told to seek Kant out by von Zedlitz on the matter of the chair at Halle. After Wolff's death the chair had gone to his follower Meier, but now that he had also died, von Zedlitz was looking for a replacement. Kant suggested one of his students. The minister, however, had his own plans.

In February 1778 Professor Kant received a letter from von Zedlitz. The latter informed Kant that although he was eighty miles away, he was attending Kant's lectures on physical geography: before him lay the notes of a student who had taken the course, and he was perusing them with interest. A week later von Zedlitz complained about the condition of the notes; the student had been slovenly: immediately after Kamchatka the notes contained descriptions of the outskirts of

Astrakhan, and there also seemed to be a mistake in the description of the fauna of Java. Kant did not pay attention to the minister's remarks (in both instances the student was not at fault, but the teacher; the mistakes were not caught and so were included in the printed text); something else caught Kant's eye — the second letter began and ended with the offer of the vacant post at Halle with the starting salary of six hundred talers.

Kant declined. (Two years earlier he had declined a post at the newly opened university at Mitau, where they had offered him eight hundred talers.) A month later von Zedlitz repeated the offer: this time the salary was eight hundred talers (Kant was drawing two hundred thirty-six talers at the time, so the offer represented a three-fold increase in remuneration!). Furthermore, the minister promised him the title of the court councillor. And last, but for Kant not least, he gave a fullsome description of the intellectual atmosphere at Halle, where the best academic powers were gathered — mathematicians, physicists, chemists, physicians, theologians.

> Look at how many good men there are at the university, which is the centre of learned Germany. And the climate is better than on the shores of the Baltic. A man who thinks as you do must admit that it is his duty to contribute to the dissemination of useful knowledge and light, that he must choose a place where his talents would be fully appreciated, where he would do the most good. I wish men of your learning and talents in your field were not so rare. I do not want to harass you, but I do not want you to misjudge your duty to be as useful as you could be in the circumstances presented to you, and I do want you to consider that there are between 1000 and 1200 students at Halle who have the right to avail themselves of your teaching. I cannot fail them.[73]

Kant would not budge. He had no need of money, renown, titles. Of course, in comparison to Halle, Königsberg was a provincial centre, and East Prussia the back yard of Europe. But this was his home; he had grown up in Königsberg, he was familiar with it, and familiarity is a second nature. Every change of life frightened him. As for his being useful, he was working so hard on his planned book precisely because he wanted to be useful. A move to another city would only interfere with his work. Thirty years before he had said that he had chosen his way. Those were then big words. He would not utter them again, but he would be guided by them. He no longer stood at the fork in the road, he had chosen and there was no going back. The *Critique of Pure Reason* had to be written!

Kant finally wrote the book in the spring and summer of 1780. Large portions had already long since been completed; it was therefore possible to finish everything in a rush, in five months. A work of fifty-five sheets lay before him (the first fourth was new). He knew the weaknesses of the book, they were largely stylistic; he did not have enough strength to rewrite and he was eager to offer the product of his labours to the judgment of the reading public.

The publisher was found easily and quickly. Johann Friedrich Hartknoch, former student of Kant's, then Kanter's employee, had opened a publishing house of his own, in Riga. In September 1780 he heard from Hamann that Kant had finished the *Critique*. Hartknoch assured the philosopher that he was not likely to find a better publisher, that everything would be done more quickly and with more taste than anywhere else; he offered the best guarantee that the book would be distributed throughout Germany. Kant gave him the manuscript. Hartknoch did not have the facilities to set type; the *Critique of Pure Reason* was printed at Halle. At first Kant wanted to dedicate the book to Lambert, who had shown such lively interest in Kant's theoretical researches. Lambert, however, had died two years before. In March 1781 Kant wrote a dedication to von Zedlitz. The book came out in May.

On the first page of the *Critique of Pure Reason* the reader encounters a motto from Francis Bacon — a lengthy quotation from the preface to *Instauratio magna scientiarum*: "We are silent about ourselves. But in the matter we are about to undertake we wish it to be taken not as a display of opinions but an honest effort to establish principles not of a sect or an ideology but of man's welfare and dignity . . . Ultimately, our *instauratio* should awaken the gentle hope that it bespeaks not the endless and the superhuman but the appropriate end and the fitting limit of endless deception."

One has not yet begun to read the book, and the words of the great reformer of philosophy already strike a very definite chord. Bacon aimed his critique at the scholastic reason and wordly sense: he demanded that dead dogmas and ossified judgments be cast aside and that all propositions which make truth claims be subjected to the test of experience ("Veritas filia temporis, non auctoritatis"). That was the proper beginning, Kant thought, and it was his task to push on from there. His guiding star was not Plato, nor Aristotle, nor his favourite Rousseau, or Hume, but the materialist and empiricist Bacon. Bacon was going to cast light on Kant's first steps across the threshold of this labyrinth of the mind.

Bacon was deeply convinced that he was going to establish measurable limits to the reign of error and in his work present a system of science which would solve the basic problems of epistemology. He died before he could complete what he had begun. The following century and a half, however, demonstrated the boldness of the attempt to circumscribe and limit the wanderings of reason. The desired end became a beginning.

Did Kant really believe that it was up to him to solve the problem Bacon had posed? Twelve years later he would write *The End of All Things* (1794) and smile ironically at man's ambition to limit what was by nature limitless. There is no need to look far and hard for an answer. The reader opens to the following pages of the *Critique of Pure Reason* and begins to understand that the answer to these complicated questions is not a mere yes or no. Human reason, Kant writes, contains within itself a strange fate: it is beset by questions which it cannot avoid because they are part of its unique nature; at the same time it cannot provide answers to them because they overtax its capacities. Reason finds itself in this predicament through no fault of its own. It begins with principles drawn from experience, but as soon as it reaches the heights of cognition it notices that new questions spring before it and that to them it has no answer. Reason is then forced to flee to new principles, which appear to be evident, yet go beyond experience. And here it is brought to a halt by contradictions which indicate an error; it cannot be brought to light by experience.

Thus begins the foreword. Kant sees as his task the overcoming of two extreme world views, i.e., two one-sided and false approaches to the problem of knowledge — dogmatism and scepticism. This is simultaneously the overcoming of Wolff, for he was the father of the division of all philosophical schools into sceptics and dogmatists. The former have their doubts about the nature of things, the latter consider that nature a matter of definition (dogma). (Wolff divides dogmatism into dualism and monism depending on whether one or two forms of being are presupposed. The monists are divided into materialists and idealists, the latter again into pluralists, who admit of the multitude of ideal essences, and egoists, who limit the number of essences by the circumference of the person.)

Kant suggests a third approach — in his opinion the only practicable — *viz.*, the critical approach. By that he does not mean the critique of books or philosophical systems: he means the critique of reason (*Vernunft*) itself, in its pure aspect, i.e., independent of all experience. Kant wants to study the instrument of knowledge before he puts it to use. (He wanted to learn how to swim without getting into the water, snorted Hegel — and he was right: Kant could never separate himself entirely from the process of cognition.)

Is reason ready to undertake a critique of itself? Kant does not doubt that his undertaking comes at the right moment. Philosophy, he says, is not philodoxy; the love of wisdom is not the love of opinion. It is time to declare war on all false teachings, all forms of dogmatism and scepticism. That cannot be done without a painstaking critique: "If the authorities think it behooves them to poke their noses into the work of scholars, their wise concern for the sciences and the society should take the form of promotion of such a critique for it alone makes the activity of reason possible. They should not support the ridiculous despotism of schools which raises hue and cry about public danger as soon as someone tears their cobwebs, of which the public have never taken note and the disappearance of which they would not bemoan."[74]

Kant's work represents the beginning of an important tradition in European intellectual development. From Kant onwards every advance in knowledge is conceived as reconsideration of previously accumulated theoretical treasures, which must be guarded with care, but must never be permitted to become fetishes. Let us not forget that the subtitle of Karl Marx's capital work is *A Critique of Political Economy*.

We now want to set out certain fundamental propositions of the *Critique of Pure Reason*. All our knowledge begins with experience, Kant says, but it is not limited by it. A large part of our knowledge proceeds directly from our cognitive activity and therefore has an *a priori* character, i.e., it lies outside of experience. Empirical knowledge is arbitrary and therefore contingent: *a priori* knowledge is universal and necessary.

(Kant's *a priorism* is different from the idealistic postulates regarding the innate ideas. Firstly, in that, on Kant's review, only forms of knowledge are prior to experience: contents are provided entirely by experience. Secondly, *a priori* forms are not innate: they are burdened by their own history. The real sense of Kantian *a priorism* lies in that the individual — as he exercises his cognitive faculty — disposes of determined cognitive forms which have been developed in him. Science is the repository of an even greater number of such forms. From the standpoint of origin the scope of knowledge is ultimately determined by the continuously growing experience of mankind. Furthermore, immediate experience coexists with mediated, acquired experience. This is how we see Kant's problem today.)

Next Kant draws the distinction between analytical and synthetic judgments. The former are elucidating in character, the latter enlarge our knowledge. The judgment that "all bodies are extended" is analytical because the concept of bodies already contains extension as a predicate. The judgment "it rained yesterday" is synthetic because the concept of yesterday is not bound up with the concept of rainy weather.

All empirical (experiential) judgments are synthetic. That is evident. But here is a question: how are synthetic *a priori* judgments possible? That is the principal question of the *Critique of Pure Reason*. Kant does not doubt that synthetic *a priori* judgments exist: were it not so there would be no pure, universally valid science. He is profoundly convinced that all mathematical judgments are *a priori* and synthetic. But how does one explain their origin? The same problem obtains in the natural sciences; they exist, grow, and provide ever new, universally binding knowledge. But how?

The thinker is careful in his judgments regarding philosophy, or metaphysics, as Kant calls his domain of knowledge: one must first establish whether it is possible as a science, whether it provides new knowledge which rests on universal, generally binding principles.

The principal question of the *Critique* — how is pure knowledge *a priori* possible — breaks down into three questions. How is pure mathematics possible? How are natural sciences possible? How is metaphysics possible as a science? Hence the tripartite division of the first major part of the *Critique of Pure Reason*: transcendental aesthetics, analytic, dialectic. (The second and third parts constitute the transcendental logic.)

Kant calls his philosophy transcendental because it examines the transcendence (*transcendo* — I cross over, I overcome) into a system of knowledge. More precisely, it examines how the conditions of experience are constructed by our cognitive faculty. Kant contrasts the transcendental to the transcendent which lies beyond the limits of possible experience and thus beyond knowledge.

Thus we come to one of the most important problems of Kant's thought, which arises within the first pages of the *Critique of Pure Reason:* experiential data, coming upon us from without, give us no adequate knowledge of the world around us. Only the *a priori* forms guarantee the universality of knowledge, but this is by no means a copy of things as they are. The thing as it appears to us (*phainomenon*), and the thing as it is in itself (*noumenon*) are, in principle, different from each other. In the 1770 dissertation Kant still held that *noumena* could be grasped by understanding; now he considers them beyond the reach of knowledge, transcendent. No matter how deeply we penetrate appearances, our knowledge of things remains different from the things as they are in themselves.[75] The division of the world into knowable *phainomena* — "appearances" — and the unknowable "things in themselves," shows a dangerous tendency towards agnosticism.

Kant did not consider himself an agnostic. He was learned, he loved knowledge, he believed in the progress of knowledge: "Through observation and analysis of appearances we penetrate to nature's inner recesses, and no one can tell how far this knowledge may in time extend."[76] The frontiers of experience are constantly expanding. But however much our knowledge increases, there always remains a frontier to be overcome: the horizon is always before us, even as we walk towards it.

Knowledge cannot be circumscribed. It is necessary to believe in science, but man should not think it all-powerful. Kant's theory of the thing in itself is a defense directed precisely against the groundless claims of science and the dogmatic prejudices about its immense power (what we nowadays call "scientism"). When it is misinterpreted, it can cause chaos; when it is properly understood, it leads to truth.

And what is truth? A difficult, yet in philosophy unavoidable question. Kant does not avoid it, but he is in no hurry to provide an answer. He says somewhat irritably that to put a reasonable question bespeaks a rational mind. If the question is meaningless in itself not only will it bring shame upon him who asks it: it will elicit an absurd answer and set a silly scene — one milks the ram, the other holds a sieve to catch milk.

Kant is touchy because the question of truth leaves him no peace and he knows that it is impossible to provide an unequivocal answer. Of course, one can say that truth is the correspondence of knowledge to its object, and he says that often enough; yet he knows that the definition is tautological. A correctly formulated question about truth runs as follows: how can we find universal criteria of truth for all knowledge? Kant's answer: there is no universal criterion of truth.[77]

What does that mean? Our philosopher is determined to crush scepticism, and now takes refuge in sceptical banalities. But let us not be too hasty in our judgment.

Kant laid the foundations (on occasion he only cleared the place for the foundations to be laid) of the noble philosophical edifice of dialectic. Kant's work is shot through with dialectical antinomies, even when they are not treated explicitly. When Kant set forth a proposition, he immediately saw the limits of its validity and the conditions that were responsible for them; he felt the need to go beyond the limits in order to connect "something with its other," as Hegel would say. The proposition brought forth its counter-proposition, the peculiar antithesis without which the thesis itself remains incomplete, incomprehensible, faulty. Kant himself did not always accomplish the synthetic step; sometimes his ideational instruments simply were not adequate to the task (that was not his fault, but his misfortune; it was the fault of the time into which he was born); but the problem is stated and the modern reader equipped with the latest developments in dialectical thought can provide antithetical positions at will.

The contradictory nature of Kantian definitions proceeds from the contradictory nature of the object under consideration. Herder could now and then show that Kant was inconsistent, but Goethe caught in his work something more significant— a crafty irony: Kant tries hard to convince the reader and then places his own positions in doubt.

That is also how he handles the problem of truth. Kant rejects the notion of a universal criterion only with regard to the content of knowledge. As far as its form is concerned he knows the criterium very well: it is the noncontradictoriness of a judgment. This is an important correction of the negative answer to the question of truth. The negative answer had destroyed the constructions of dogmatists; now the task is to avoid the snares of the sceptics. Kant knows that the exclusion of contradiction represents a "purely negative criterion of truth,"[78] but if one makes use of it a solid edifice of science can be erected.

We now know that science has at its disposal also a "positive" criterion of truth — praxis. Dialectical materialism introduced praxis into the theory of knowledge. Kant's criticism made that possible: it showed that in "pure" theory it is possible to formulate only the negative criterion of truth. Praxis is the compass in the hands of the scholar. We know that the criterion of praxis is not absolute because it can neither confirm nor disprove the fullness of a specific human representation.

Praxis confirmed the claims of non-Euclidean geometry; another praxis confirmed the truth of Euclid's axioms. Truth is a process whereby the world is grasped ever more completely, it is the movement from ignorance to knowledge, from incomplete knowledge to more complete knowledge, a movement that cannot end because the world is inexhaustible. Hegel, who was the first to formulate this dialectical understanding, nonetheless contradicted himself when he proposed that complete correspondence between the object and thought — absolute knowledge — was possible. (Hypergnosticism is just as dangerous as agnosticism.) Kant was more careful: his transcendental "thing in itself" (*Ding an sich*) reminds us that there are moveable limits within knowledge, but no final limits to knowledge itself.

The "thing in itself" brings another circumstance to our attention: there are areas where science is powerless. For instance, man's conduct: the realm of freedom, or, more properly, of choice (*Willkür*). Literature and the arts before and since Kant have borne witness to the simple fact that man not only does not follow the dictates of science, he even refuses to follow the rules of elementary logic. In the *Critique of Pure Reason* we read: "No one can form judgments about things relying solely on logic."[79]

All Kant's reservations and amplifications serve a noble aim: to banish phantoms from truth so that those who seek it do not fall prey to disappointment. "The land of pure understanding (*Verstand*)," the philosopher says,

> is an island, enclosed by nature itself within unalterable limits. It is the land of truth — enchanting name! — surrounded by a wide and stormy ocean, the native home of illusion, where many a fog bank and many a swiftly melting iceberg give the deceptive appearance of farther shores, deluding the adventurous seafarer ever anew with empty hopes, and engaging him

> in enterprises which he can never abandon and yet is unable to carry to completion. Before we venture on this sea, to explore it in all directions and to obtain assurance whether there be any ground for such hopes, it will be well to begin by casting a glance upon the map of the land which we are about to leave, and to enquire, first, whether we cannot in any case be satisfied with what it contains — are not, indeed, under compulsion to be satisfied, inasmuch as there may be no other territory upon which we can settle; and, secondly, by what title we possess even this domain, and can consider ourselves as secured against all opposing claims.[80]

So the philosopher spreads the map of the land of knowledge (the realm of truth) before us and suggests that it be studied with care. The most important road into the hinterlands is the road of sensuous knowledge. The corresponding part of the *Critique* is entitled the "transcendental aesthetic." In Kant's view there exist two *a priori*, pre-experiential forms of sensibility — space and time. Space systematises sensations from without, time those from within. Bertrand Russell offers the following elucidation of Kant's thought: if you wear blue glasses, the world appears blue: so man looks at the world through the special glasses of space and sees everything in spatial relationships.

Kant did not deny the empirical reality of space and time. He thought that the doctrine of the ideality of space and time was at the same time the doctrine of the thoroughgoing reality of both in relation to the sense organs. Kant never gave up his cosmogonic hypothesis: the real processes of birth and death of worlds take place in real space. He did not believe that the world would end when all time ran out. In the *Critique of Pure Reason* he pursued an epistemological problem — whence come our representations of duration and expansion? He was convinced that they could not be drawn from experience; they are *a priori* and consequently universal and necessary. Only they made the science of quantity — mathematics — possible. However, *a priori* does not mean innate. How then do the representations of space and time arise? Kant answers that they are created by the imagination (*Einbildungskraft*).

Kant's views of space and time were to a certain degree a reaction against the mechanistic notions of absolute duration and expansion as two independently existing, empty containers for things. Kant considers space and time in their reciprocal relationship, but that relationship is established only in the knowing subject. In the world of things in themselves, beyond man there are other modes of coexistence and things following upon one another. Today we know of the existence of non-Euclidian geometry. We also know that our optical perception differs from the world offered to it.

Time plays an important role in Kant's philosophical constructs. His interpretation of time can only be called poetic. It is not accidental that two Russian poets were inspired by Kant's transcendental aesthetics.

Andrey Bely:

> In spring the gaze looks out into the distance:
> The azure depths are there . . .
> But *Critiques* are before me —
> Their leather bindings . . .
> In the distances — the starry-eyed attire
> Of another being . . .
> And, with a start, I remember
> The illusory nature of space.[81]

Alexander Blok:

> I sit behind a screen. My legs
> Are so fragile . . .
> My hands are so small
> And the window is so dark.
> It is warm here and dark. I blow out
> The candle that someone has brought
> But show gratitide . . .
> They have been begging me to amuse myself,
> But these small hands . . . I am in love
> With my wrinkled skin . . .
> I can live a lovely dream,
> But I will not take the trouble:
> I will not bother to disturb my dream
> Of these lights in the window . . .
> I cross my small hands,
> I cross my fragile legs.
> I sit behind a screen. It is warm.
> Someone is here. I need no light.
> The bottomless eyes, like glass.
> On the wrinkled hand — small rings.[82]

The poem is entitled "Immanuel Kant." The poet claims to have been inspired by the transcendental aesthetic. There in dark oblivion, sunk in thought, aloof, sits a tiny, fragile man. So the poet sees the philosopher. What a grotesque image! Where are time and space? Perhaps Blok read this part of the work by chance (as he could have read any other part) and expressed in verse his relationship to Kant's personality? Is transcendental aesthetic just a subterfuge? It is not impossible. But it is more likely that something else is going on; we will return to it later.

The indubitable achievement of Kant's epistemology is a new view of the inter-relationship between intuition (*Anschauung*) and intellect. In the seventeenth century there were two opposing theories of cognition — sensualism and rationalism. Sensualists claimed that sensual knowledge had pride of place; rationalists, naturally, gave that place to intellect. But neither school saw the difference, in principle, between the two modes of cognition. For the sensualists logical cognition was merely the perfected form of perception (Locke believed that there was nothing in the understanding that had not first been in the sense); rationalists, on the other hand, saw sense perception as a kind of potential reason. Kant emphasized that neither "cognitive stem" could be re duced to the other: "Neither of these capacities should be preferred to the other. Without sensibility no object would be given to us, without understanding no object would be thought. Thoughts without content are empty, intuitions without concepts are blind."[83] Scientific knowledge is the synthesis of sensibility (*Sinnlichkeit*) and understanding.

How is that synthesis accomplished? The answer takes up a large portion of the transcendental logic. The traditional formal logic saw its task to be the investigation of the structure of abstract thought, of thought separated from its content. Kant proposes a reform of logic: "We should have a logic in which we do not abstract from the entire content of knowledge . . . it would also treat of the origin of the modes in which we know objects, in so far as that origin cannot be attributed to the objects."[84] The last proviso is very essential. But we are interested here not in the proviso but in the issue: Kant's transcendental logic is contentual, it enquires into the origins, scope, and meaning of knowledge. And another thing must be clearly realized: Kant speaks a lot about form, but he is thinking of a form related to content; for him there is no empty, content-free form. Kant is not a formalist (this last point is very important for a correct understanding of his philosophy of art).

Now to continue with the synthesis of knowledge. Kant calls the logical forms which serve this fundamental synthesis categories, as did Aristotle. Categories, in Kant's view, are *a priori*. (They are however not innate; we develop them in the course of "epigenesis of pure reason".) Corresponding to the four different kinds of judgment, we find in Kant the following table of categories: 1. Category of quantity: Unity. Plurality. Totality. 2. Category of quality: Reality. Negation. Limitation. 3. Category of relation: The inherent and the subsistent (*substantia et accidens*). Causality and dependence. Community (agency and passivity). 4. Category of modality: Possibility — impossibility. Being (*Dasein*) — Non-being (*Nichtsein*). Necessity — Contingency.

The tripartite division of each group of categories is striking. Here it is already possible to see clearly Hegel's triad — thesis, antithesis, synthesis. Later (in the *Critique of Judgment*) Kant clarifies his thought with the following statement:

> It has been thought somewhat suspicious that my divisions in pure philosophy should almost always come out tripartite. But it is due to the nature of the case. If a division is to be *a priori*. . . the division must be tripartite because that is the requirement of synthetic unity, namely there is, first, a condition, second, a conditioned and third, the concept arising from the union of the conditioned with its condition.[85]

Categories are very general concepts, the skeleton of cognition, as it were. Only thanks to their existence is "pure" natural science possible. But just as it is not enough to draw skin over a skeleton to get an organism, so the body of science is something considerably more complex than the twelve categorical forms stuffed with empirical content. From each category, categories of lesser universality are deduced. The category of causality, for example, is filled out by the concepts of force, activity, passivity; the category of community by presence, opposition, etc. Kant says that if he wanted to he could show the whole "genealogical tree of pure understanding";[86] he does not do that, in order not to digress; his task is not to show the completeness of the whole system, but the completeness of all of the principles grounding it.

To a dialectical materialist all philosophical categories represent abstractions of the real connexions between objective activity. Kant sees it differently: "We can imagine nothing that is bound in the object until we have bound it ourselves." There is a more decisive clarification: "Understanding (*Verstand*) does not draw its laws (*a priori*) from nature, it ordains them for nature."[87]

Kant's phrase requires commentary. If we misunderstand it, we cannot grasp the major point of his epistemology — the idea of the activity of our consciousness: and Kant saw his major contribution precisely in this. He even compared himself to Copernicus, in the belief that his revolution in philosophy was no less profound: before him it was thought that our cognition must direct itself to the objects. Kant's starting point is that objects must direct themselves to our cognition.

This sounds paradoxical, but we should not fear paradoxes: they awaken thought and force it to new paths. The new on which Kant insisted was the insight into the active role of consciousness. All of pre-Kantian philosophy viewed intellect as the passive repository of ideas, which find their way there by natural or supernatural means. Idealists were as wrong in that regard as materialists. Kant went too far with his categorical formulations, but it was for a good cause: for in order to straighten the crooked, man must bend it in the other direction. And epistemology before Kant was full of crooked twigs, and all of them were bent to the same side.

Furthermore, we know that when Kant says "the object of knowledge" (even "nature") he does not mean the thing in itself but its appearances, that is, that part of reality which interacts with our cognition. And that part of reality — and this cannot be disregarded — corresponds to acts we implement in the course of our understanding; the realm of reality of the appearances is contained within the system of our rationally organised experience. Marx respected the classical German philosophy (Kant was its originator) precisely because it emphasised the active aspect of cognition. Consciousness does not merely mirror the world, it makes it — that is an axiom of dialectical materialism. Its roots are in Kant.

The issue here is not so much the individual consciousness of a solitary man as it is "consciousness in general," the totality of social intellectual achievements embodied in language and other forms of culture. Transcendental consciousness, which, in Kant's view, determines "nature," is hence a complete, structured map of the world. The transcendental subject is humanity as much as an individual human being.

Matters do not stop there. Kant's doctrine of the activity of consciousness threw light on a most mysterious process — the formation of concepts (*Begriffsbildung*). Kant's great predecessors who tried to solve this problem always ended up in blind alleys. The sensualists belaboured induction, the derivation of certain universal characteristics and principles from facts of experience. We know from experience that swans are white, and crows black. But even here common sense is very sceptical about such generalisations: the expression "white crow" refers to an extremely rare instance which nevertheless is possible in the usual order of things; and black swans do exist. And how can we use induction, the process of abstraction of universal characteristics, to explain instances of intellectual creation and the construction of something entirely new — a machine or a scientific theory?

The rationalists followed another path. They observed the strict correspondences, independent of men, between the order of things and the order of ideas. They considered thinking an "automaton spirituale" (Spinoza's term) which prints out truth by following a previously given, "pre-established" (Leibniz term) programme. The explanation was sound, but it contained an essential error: it could not explain mistakes. Descartes' attempt to solve the problem was characteristic. He saw the root of mistakes in free will: the less man darkens the light of divine truth, the more is he protected from error; passivity, then, is the guarantee of correct knowledge.

Kant, like Copernicus, breaks radically with tradition. He sees in the human intellect a construct erected in advance — categories, but this is not yet scientific knowledge, only the possibility of it. Empirical data constitute a similar possibility — they are building blocks which have to be put into the framework of categories. The building, howevery, requires a builder; Kant gives him the name of productive imagination (*Einbildungskraft*).

Until Kant, imagination was held to be the prerogative of poets. The dry pedant from Königsberg also saw a poetic principle in science: in the act of concept formation. A man who lived like an automaton rejected the appelation of automatism for human intellect. On Kant's view spirit is a free creator.

Perhaps Kant's love of geography helped him to envisage the role of imagination in cognition: he used to tell students of countries he had never seen but had imagined from accounts of others. Kant could have read some essential points in Voltaire's article on imagination, written for the famous *Encyclopédie.* Imagination is present in applied mathemathics, claimed Voltaire; Archimedes' imagination was as strongly developed as Homer's!

What is the verdict of contemporary scientists? I have before me the work of the well-known French mathematician Jacques Solomon Hadamard, *The Psychology of Invention in the Mathematical Field.* There is an essay in the book by an even better known mathematician, Henri Poincaré. Hadamard did not conceal the source of his thought. The point at issue is that the process of scientific creation can be broken down into four stages — preparation, incubation, enlightenment, conclusion. The first and the fourth stages are dependent on the activities of cognition, the second and the third on the unknown components of the psyche. Thus scientific discovery is bound up in its middle stage with the uncontrolled work of the intellect. Enlightenment occurs when the conscious components of thinking diminish, or cease completely.

This does not mean that discovery is accidental by nature. Only the highly developed consciousness of the scholar can create the "heuristic situation," fix its attention on new data and see how they contradict the existing theory. Only consciousness can set the problem and give a general direction to research. But what happens then? An entirely new mode of thinking becomes engaged, not the usual thinking connected with a fixed notational system necessary for communication. Creative thinking requires more supple forms. "When I think, words are totally absent from my mind," claims Hadamard; instead of words he then uses "makings of indistinct form." Henri Poincaré, apologising for the roughness of comparison, suggests that the components of the unconscious be thought of as "atoms" which before the inception of intellectual labour exist in the state of dormancy, as if "glued to the wall." The first conscious labour, when attention is directed at the problem, sets these atoms in motion. Then consciousness can relax, the unconscious process of thinking only appears to relax because the work of the "atoms" of the sub-conscious does not cease until the solution has been found. "After the receipt of the impulse from our will atoms do not return to their original state of immobility. They dance on on their own." Creation is the making of choice and the discarding of unnecessary variants; the feeling of scientific beauty guides unconscious choices.[88]

But let us return to Kant. He does not use the term "the unconscious" (*"das Unbewußte"*). In the previous chapter we begged the reader to remember another term: "dark representations," which he used in the 1764 submission to the Berlin Academy of Sciences. Here is an eloquent reflection from his *opus postumus*: "Understanding (*Verstand*) works largely in the dark . . . Dark representations are pregnant with clear ones. Lesson. It is only necessary to bring them to clarity. The midwife of thought. All acts of the understanding (*Verstand*) and reason (*Vernunft*) can be born in darkness . . . Beauty must be ineffable. We cannot always say what we have thought out."[89]

If one adds to this another series of analogous formulations, just as decisive as this one, which are scattered throughout Kant's notes, then the identity of his views with the views of Hadamard and Poincaré fills one with wonder. Modern heuristics deals with problems which had already preoccupied Kant.

Perhaps we should say that in the *Critique of Pure Reason* the term "dark representations" does not appear (Kant discusses them in earlier and later works). Nevertheless the idea of an unconscious, yet active creative principle is stated in clear, unambiguous terms. Kant speaks of the spontaneity of thinking. Thanks to the productive imagination, understanding itself is spontaneous, i.e., it develops its concepts in an elemental fashion, without the control of consciousness. "Insofar as imagination is spontaneity, I sometimes also call it productive imagination."[90] This is one of the central ideas of the critique of pure reason.

But where is the guarantee, the reader might ask, that our imagination brings forth the desired concept, and not God knows what? That is the problem of true knowledge, of which we have already spoken. Besides, as we have already explained, productive imagination is determined, first, by ready-made constructs (categories), and second, by the given building material — empirical data. That is precisely why imagination does not produce castles in the air but substantive scientific edifices. Productive imagination is not empty fantasy. It is the instrument of the synthesis of sense and understanding.

Kant provides few illustrations of this general statement of fact. He wishes to define with precision the progress of this synthesis and discovers an intermediate phase in it, a link between sensual activity and abstract thought. In the *Critique of Pure Reason* a new term appears: "the schema." It is the semi-finished product of productive imagination, something quite wonderful — on the one side sensual, on the other intellectual: a "mediating representation," a "sensual concept." Hegal will (in another context) call it the "abstract representation;" but the great dialectician is not interested in the problem of such a synthesis at all: from the sensual he will expect only trouble; spirit, in his opinion, follows its path to truth without sensual material, the truth which is the identity of being and thought. Kant is not quite as well known as a dialectician, but he is the one who seeks solutions to the dialectical problem: the unification of the opposites of the sensual and the logical.

The schema, Kant emphasises, must be differentiated from the picture. The latter is always phenomenally accessible. Set five dots next to one another and you have the picture of a definite quantity; the pure schema of quantity, however, is number alone. Schemata, not representations, are thus the ground of concepts. How they come forth, it is difficult to tell: it is "an art hidden in the depths of the human soul, an art the true nature of whose deeds we will some time divine and reveal to the eye." It is true that Kant points to the mediating mechanism of the synthesis of sense and understanding — time. Temporal succession appertains equally to perception and concepts. Time is the foundation of the schema.

If we cast another glance at the categories it becomes clear how Kant sees the genesis of categories, which precedes all experience: we have already encountered the "epigenesis of pure reason."[91] Each category has its own schema. What, for instance, is the schema of substance? The permanence of the real in time. Or the schema of cause? A succession of realities. With the help of time our imagination constructs the categories, even the categories which are ordained in nature. In order to break through the boundaries of nature, we must overcome time. We must remember this conclusion; we shall need it when we return to Blok.

Now the special nature of intuition does not explain this conclusion, although this is the place for the discussion of the concept of intuition. It appears that Kant cast intuition out of the sphere of intellect and placed it fully in the sphere of the sensual. Actually Kant only threw out the word, the notion itself he handled thoroughly and richly. Descartes, Spinoza, and their predecessors used the term "intuition" to designate a passive immediate contemplation of truth. Kant accepts this interpretation. In so far as intellect is active, then, there is no room in it for intuition. Intuition (*Anschauung*), in Kant's view, can only be sensuous.

For us intuition is something different. We see in it an immediate capacity for creation, the discovery of what we need. There is no passive cognition. On the evidence of recent research in psychology, the human psyche is ever active on the level of sense organs; there is no automatic photography here but a unique creative process. The active principle in intellect, which Kant calls productive imagination, is nothing other than the play of intuition.

Intuition is instrumental not only in the development of concepts, but in their application as well. The scholar (and everybody else) must be in possession of an instrumentarium of universal rules, laws, principles; he must also know how to use them in concrete, individual instances. Kant calls this intuitive capacity the "facility of judgment" (*Urteilsvermögen*). Judgment is "also the specific ingredient of the so-called native wit, whose lack no schooling can remedy; although schooling gives us a limited number of the rules of understanding, borrowed from another's insight, the capacity to use them properly must be present in the pupil himself, and no rules that man can devise to accomplish this aim can ensure him from misuse . . . The absence of judgment is actually what we call stupidity, and there is no cure for this infirmity."[92]

Kant has described intuition in other ways, closer to our modern understanding (later on we shall deal with artistic intuition and moral intuition). Intuition accompanies cognition in all its movements: "upwards," towards abstraction, and "downwards," towards the connexions of these abstractions with the thing. In the first instance productive imagination is at work, in the second, judgment. Understanding cannot function without both of them.

So we have understanding bound up with intuitive judgment. Besides them Kant defines yet another sphere of intellectual activity, its highest stage — reason (*Vernunft*). In the broadest sense of the term, reason is equivalent for Kant to all logical thought. Sometimes he is not quite consistent and uses the term "understanding" (*Verstand*) in this sense. In the narrowest sense understanding pertains to the sphere of science: here the cognitive synthesis is realized; reason, on the other hand, is the highest controlling and regulating instance: its sphere is philosophy. The Analytic is the doctrine of the understanding, the Dialectic that of reason.

It was with the Analytic that Kant had struggled for years. Three fourths of the book were already written, but he had no beginning, and no central section. He could not solve the principal issue: how scientific knowledge came to be. He stood before the problem as if before a wall, trying to dislodge a few bricks with his fingers. With an iron determination he stuck to his task, day after day, month after month. And the wall gave way. It came like a flash of lighting — that would remain with us forever.

Progress in science is the liberation from prejudices. One of them is that science is all-powerful. That was an illusion of the early Enlightenment: science can do anything; it can prove the existence of God, ground the immortality of soul, reveal all the secrets of man's being. One of the tasks of Kant's self-critique on reason was to show that such pretensions of reason are groundless.

Dialectic, in Kant's view, is the logic of appearance. For reason is capable of creating illusions, of mistaking the apparent for the actual. The task of critique is to bring clarity to this confusion. That is why Kant's *Dialectic* begins with the definition of the concept "appearance" (*Schein*). It is not a hallucination, not a phantom which would be a product of individual cognition. Appearance is an "unavoidable" illusion.[93] It appears to all of us that the sun travels across the sky. But if we wish to grasp the natural appearance of day and night we are forced to accept the explanation that the earth revolves around its own axis; illusion and appearance are different concepts. And if man does not pay attention to the rules of logic, empirical appearance can be joined by logical appearance. This error, however, is easily spotted and corrected. With the philosophical, transcendental appearance, matters are considerably more difficult, since judgments here entail things which lie beyond the limits of possible experience. So it is, for instance, in the judgment: the world must have a beginning in time.

For reason the difficulty lies in that it here does not deal with scientific concepts (the sphere of the understanding) but with ideas. The Idea (*Idee*) is a concept for which it is impossible to proffer an adequate object. In its immediate activity reason is not directed to experience but to the understanding; reason must first and foremost delineate for understanding its field of activity. Reason works out fundamental principles (*Grundsätze*) and universal principles (*Prinzipien*) for the understanding and judgment to apply in specific instances. Reason fulfills a regulative function in the cognitive process, it points understanding towards determinate goals and sets goals for it. (The function of understanding is constitutive, that is to say, constructive: it creates concepts.) Reason purifies and systematises cognition.

Therefore it is wrong to think that Kant slights reason. In a properly disciplined approach this organ of thought performs a worthy service. For it is only thanks to reason that theory transforms itself into praxis; its ideas regulate not only our cognition but also our conduct. Just because man will never act totally in accordance with an idea of pure virtue, it does not follow that such an idea is a chimera. All moral judgments are possible only by means of such ideas.

Kant reminds us of Plato's idea of a perfect state. The fact that it is not realisable should not deter us from trying to embody it. A government based on the highest degree of human freedom, where the freedom of one is identical with the freedom of all, is a necessary idea which informs the basis of every new system of laws. Nothing is more harmful and less worthy of a philosopher than to capitulate ignorantly to given contradictory experience. And even if, in Kant's view, a perfect order will never come to be, the true idea which postulates these principles in the form of a prototype, as it were, still remains so that we can gradually bring the general social order to a more complete state of perfection by their guidance.

Ideas play a great role in the theoretical sphere as well. Reason points up the limitation of the categorical synthesis of understanding, while it — as reason — can achieve the highest degree of generalization, beyond the frontiers of experience. Theoretical ideas build a system drawn from the three possible variants of the concept-reality relationship: first, the relationship between concept and subject, second, the relationship between concept and object, third, the relationship between concept and both subject and object, namely the relationship as it embraces all things. Thus we get three classes of ideas — ideas of soul, ideas of the world, and ideas of God.

Pre-Kantian philosophy made judgments concerning all three classes with frivolous ease. Kant thinks that in this area in particular reason should subject itself to self-control and self-criticism. What is the value of those confusing judgments about the soul or substance and the ensuing propositions regarding immortality? Kant calls such judgments — paralogisms, formally false rational conclusions. He refutes them without difficulty.

In the second edition of the *Critique of Pure Reason* (1787) Kant's sharp critique had a very specific target. A new section appeared in the text: "A Refutation of Mendelssohnian Proofs of the Existence of the Soul." In his *Phaidon*, Moses Mendelssohn had grounded the immortality of soul on its simplicity and indivisibility. That which is indivisible cannot diminish, consequently it cannot disappear. Kant remarked with respect to this proof that diminution can take the form of not only a change in size, i.e., extensive change, but also a change in the degree of reality, i.e., intensive change. Consciousness cannot be broken down to atoms, but it has different degrees of reality.

Two years before the second edition of the *Critique of Pure Reason* an anonymous work, *Anti-Phaidon*, appeared on the German literary market. Its author, Karl Spazier, a teacher at the *Philanthropin* at Dessau, man of letters and philosopher, heralded Kant's critical arguments against Mendelssohn. Spazier remarked that the intensity of consciousness changed in relation to age: it grew in youth, it diminished in old age. It is difficult to tell whether Kant knew of the *Anti-Phaidon*.

But Spazier knew Kant's works. In his book he calls Kant one of the "greatest thinkers" and refers to *Dreams of a Ghost-Seer* and the *Critique of Pure Reason*. This circumstance is very important: Kant's critique destroyed idealistic dogmas and thereby played an important role in the establishment of a materialist understanding of the world. Spazier was a materialist of an unusual sort: he thought that soul was material and therefore immortal. Such range of materialistic understandings was widespread in Germany: the German wanted to stand at the pinnacle of science, but he did not want to quarrel with God. Both lines of thought are present in Kant's work as well. He was, however, wary of jumping to hasty conclusions.

Now to the next group of concepts of pure reason — the cosmological ideas. First we should acquaint the reader with the import of the problem. Just as productive imagination represents the culminating point of the analytic, the antinomies are the high point of Kant's dialectic. This is the second important complex of problems of the *Critique of Pure Reason*. Much has been written about the antinomies for this reason, much which is profound and even beautiful. It is easy to find examples. The following excerpt from the book *Dostoevsky and Kant* by the Soviet literary scholar Y.E. Golosovker, is presented as a background to a more fundamental discussion of the problem.

Whatever philosophical road a thinker decides to follow he must cross a bridge — Kant. Although this bridge — one of the seven wonders of the world of speculative construction — is protected by the high dykes of human experience, an icy wind of hopelessness pierces the traveller through and through, and in vain will he look for the sun of life in the surrounding gloom. And no matter how slowly and carefully our chilled thinker makes his way, with frequent stops, before he has reached the middle of the bridge he will have noticed that his step is becoming unsure, that the bridge under him shakes and shivers, that he is upon a very suspiciously sceptical road . . .

And suddenly the bridge begins to toss our thinker from one side to the other; he drops towards the bottom, then he flies up, into the sky, as if the bridge were not a bridge but a swiftly moving system of dancing see-saws which toss him here and there, from one arm to the other. Moreover, among these see-saws gone berserk he notices a yawning abyss whose bottom his eye cannot reach. And if the need of stability, the essential element of culture, prompts him, the confirmed dogmatist, to seek firm ground under his feet, a yes or a no, right in the midst of these four dancing see-saws (yes, there are four of them), if this dogmatic necessity subdues the instinct to speedily retire from this world of centipedal sophistic sections — then is the wanderer lost. He will never leave the region, he is doomed to spend the rest of his days on these see-saws, slipping from one arm onto another, becoming a pendulum himself until madness or death deliver him from this spiritual torture. But if the wanderer has managed to cross the devil's bridge to the other end, then it is worth the trouble to turn around and behold what he has accomplished: he will then throw up his arms and berate himself for having taken the grandiose joke of the builder seriously: for behind him lies an illusory actuality, the product of his own dogmatic stubbornness which hid from him an indefatigable scepticism.[94]

Now to the matter at hand. Antinomies are mutually exclusive judgments both of which have the same power of proof; reason inevitably encounters antinomies when it seeks to understand the world as a whole. In concordance with the four-partite table of categories four cosmological ideas offer themselves to reason: in them, theses and antitheses are equally well substantiated. (If Kant's tripartite division of categories corresponds to Hegel's triadic structure of being, then the antinomies are heralds of mutually permeating "coupled" categories of Hegel's theory of being.)

First thesis: The world has a beginning (a limit) in time and space. Antithesis: The world is spatially and temporally infinite. Second thesis: Everything in the world consists of simple elements. Antithesis: There are no simple elements, only complex structures. Third thesis: Causality exists in the world through freedom. Antithesis: There is no freedom, everything happens according to nature. Fourth thesis: In the world of causes there exists an absolutely necessary being. Antithesis: In that world there is nothing necessary, everything in it is contingent.

Let us look at the first antinomy more closely: the proofs are given in such a way that the impossibility of both thesis and antithesis is demonstrated. When Kant proves the thesis of the first antinomy, he presupposes that the world is not limited by time. Infinite time, however, proves to be limited in each given instance (the present) and that contradicts the idea of infinity; the infinite past which is at the same time a "successive" world order is impossible. When Kant considers the antithesis he again begins with the opposing proposition: the world is limited in time; then, however, we must accept the existence of an "empty" time, in which the world did not exist. Again the result is nonsense.

Does reason want to batter down the wall with its head? Better not. It will not receive any lump on the head, but a slight mental confusion cannot be ruled out. For the wall is an illusion.

If someone says that an object either smells good or bad, it is possible to suggest a third possibility: the given object doesn't smell at all, both propositions are false. Such is the case in the instance above: in spite of an apparent foundation in fact, both propositions, regarding the finitude and the infinity of the world, are false. They pertain to the world of appearances; in the world of things in themselves there is a third possibility. Kant formulates the dialectical problem. "Here we see the strangest phenomenon of human reason, another example which cannot be found in any application of reason."[95] When reason strives to grasp the world as a whole and reveal its essence, it inevitably runs into contradictions. Contradiction is an inescapable moment of thought.

Kant deals with the second antinomy in an analogous fashion. Here, again, he considers both the thesis and the antithesis to be equally false. The third and the fourth antinomies are different: in them both the theses and the antitheses are true.

The third antinomy is the most important one for Kant. There is no freedom, everything in the world is rigidly determined. Yes, that is truly the case in the world of appearances. But man is given a free will, and natural determination has no power over him. This is equally true, Kant maintains, and offers a further clarification: but only in the world of things in themselves.

Man lives in two worlds. On the one hand he is a phenomenon, a cell of the sensible world which exists according to its laws, often far from the spirit of humanity. On the other hand, he is a noumenon, a supersensible being, subjected to the ideal. Man has two characters: one is empirical, determined by the world around him, the other noumenal, intelligible, in which his own inner nature is given to him. Are they connected? Or is the intelligible character something beyond this world, something that cannot become apparent in the world given to us? No: in man's mortal conduct the connexion between the two characters is realised. That is the foundation of man's responsibility.

Kant clarifies his thought with a simple example. When man lies wickedly and his lies produce an upheaval in society, who is to blame? If we are guided by man's empirical character we can find causes of such evil behaviour in poor upbringing or bad environment. But even when we ascribe the transgression to such causes we do not exonerate the evil-doer because he was subject to their influence: we contemplate the deed itself and point to reason as the cause that could have and should have shaped his behaviour differently. "Behaviour is ascribed to man's intelligible character: when he lies he is totally responsible; regardless of all the empirical causes of behaviour, reason was completely free, and the deed is to be regarded as a failure of reason."[96]

Here we have reached a high point of the critical philosophy and at the same time the cause of its coming to be. The later Kant, remembering the history of the *Critique of Pure Reason*, wrote in a letter that it was precisely the problem of freedom ("Man is free and, conversely, there is no freedom, everything is natural necessity") that shook him out of his dogmatic slumbers and made him subject reason to a critique so that he would eliminate its scandalously self-contradictory nature.[97] Thus the first question of the *Critique* — how are synthetic *a priori* judgments possible — rose together with the second, for Kant a more important question — how is human freedom possible? There is freedom, but where is it? We cannot discover it in the world of "appearances." Man is free only in the world of "things in themselves." Kant's dualism is a peculiar attempt to justify man's necessarily ambiguous behaviour in an antagonistic society where adaptation to circumstances is the law, and moral behaviour an act of heroism.

Kant also introduced the ideality of time, which so thoroughly shocked materialist thinkers, only to accomplish his purpose — to save freedom. Time orders appearances in nature, a genetic linkage of states immune to human intervention. Freedom, however, requires that man have power over himself. Therefore, if time is a predicate of things in themselves, freedom is impossible. Only because there is no iron link between causes and effects in the intelligible world, that is to say because there is no time, can unique causality exist there: "through freedom" — and only it makes man a moral being. In nature one and the same cause always produces one and the same effect — with an iron necessity. Man however can foresee the consequences and in unchanging circumstances act differently. He can eliminate time, effect its progress. So understood, time's physical reality is transformed into a poetic metaphor.

It is now a little clearer why Blok wrote his poem dedicated to Kant on the transcendental aesthetic, namely that part of the *Critique* which treats of space and time. Blok considered that chapter to be the most important part of Kant's philosophy. "There are, one could say, two kinds of time and two kinds of space: one is historical, calendar-like, the other uncountable, musical."[98] "Musicality" for Blok is the ability to partake in the harmony of the world, to break through the barriers of historical determination and civilisation and escape them. That is how he saw Kant in his poem.

"What is time? There is no time; time is numbers." This is written in the manuscript of *Crime and Punishment.*[99] Dostoevsky was familiar with the *Critique of Pure Reason.* (When he was in Siberia he asked that a French edition of the book be sent to him.) Did he have the book before him when he pondered the novel about Rodion Raskolnikov, who acted in defiance of the norms of morality and law? Or when he wrote the life-story of prince Myshkin? Or that of Aleksey Fyodorovich Karamazov? Well, it is hard to tell. But the *Dostoevsky and Kant* theme has the right to exist.

Golosovker's book bears the subtitle: "Meditations of a reader about *The Brothers Karamazov* and the *Critique of Pure Reason*." The philosopher's name and the title of his book do not appear in the pages of the novel. But there is rich material for a comparison. Golosovker states: "Dostoevsky not only knew the antithetic of the *Critique of Pure Reason*, he had thought it through. Even more: in the motivations of dramatic situations of the novel he took it into consideration. He made Kant, or, more precisely, the antithesis of his antinomies the symbol of everything he fought (in himself and in his opponents) as a writer and a thinker." At issue here is the fourth antinomy: there is "an absolutely necessary being" (as Kant designates God); nothing is absolutely necessary, God cannot be found in the world of nature. His existence (*Dasein*) can only by thought of in the realm of freedom, in the world of things in themselves. "A mortal conflict" arose between Kant and Dostoevsky, Golosovker claims, "one of the most profound conflicts in the history of human thought."[100] Nicely put, but an essential clarification is needed: the conflict is purely an appearance. Kant and Dostoevsky are not antipodes, they strive towards the same goal — the freedom of human personality.

Dostoevsky must have found the antithesis of the fourth antinomy, the factual denial of God's existence, a deadly enemy; for if "there is no God, then everything is possible," "murder by proxy" guides Smerdyakov's hand. But Kant did not like this antithesis either. The endless shuttling between thesis and antithesis brings to mind Ivan Karamazov, thesis and antithesis torture the Dostoevsky of *The Brothers Karamazov*; the author of the *Critique of Pure Reason* is also disturbed by them, as we see from Kant's notes. Kant knows one thing well: faith and knowledge must not be mistaken for one another. The world of nature is accessible to knowledge, the world of freedom — well, that is another story.

Men who had little use for faith rose against Kant: they wanted to know God just as they wanted to know something about their monarch. Heine compared Kant to Robespierre: the French revolutionary made an attempt on the king's life, the Königsberg philosopher dared to do more — he lifted his hand against God.

> People really honour Robespierre too highly when they compare him to Immanuel Kant. Robespierre, the great bourgeois of rue Saint-Honoré, was subject to the furious mania for destruction when it came to kingly power and he then gave free

rein to his regicidal epilepsy; but as soon as it came to the highest being he wiped the foam from his mouth and blood from his hands, put on his blue Sunday coat with shining buttons and even stuck a sprig of flowers in the wide lapel . . .

But if Immanuel Kant, the great destroyer in the realm of thought, greatly surpassed the terrorism of Robespierre, he did have a few things in common with him which allow a comparison of the two. First, we find in both of them an inexorable, sharp, prosaic, sober honesty. Next we find the same talent for doubt: only the one exercises it with respect to thought and calls it critique, while the other turns that talent against the people and calls it republican virtue. The bourgeois reveals itself in both of them to the highest degree — both were meant by nature to weigh sugar and coffee, but fate decreed otherwise and put by the one the king and by the other God in the scales . . . And they gave the proper weight![101]

It is easy to imagine how the *Critique of Pure Reason* must have shaken the minds of the philosopher's contemporaries when we see that three decades after Kant's death so radical a man as Heinrich Heine heaped scorn upon him! It is unfair to chastise Kant as bourgeois; such a judgment only serves to reveal the profoundly bourgeois nature of Heine himself — the nature which prevented him from seeing and recognizing Kant's great creation in "the realm of thought." The poet confesses that whenever he listens to someone arguing about the existence of God he is seized by a strange anxiety, an eery anguish of the kind he first experienced in an asylum when, surrounded by the inmates, he lost sight of his guide. The prospect of being alone, without a guide in this mad world terrifies Heine. But to measure his forces against Kant in the defense of the almighty was beyond his powers. He could only express his dissatisfaction and take pleasure in catching the great annihilator of dogmas in a contradiction.

Only Hegel could enter into an open disagreement with Kant: Hegel worked very hard in his lectures *On the Proof of God's Being* to restore what had been destroyed in Lutheran theology under the blows of critical reason. For Kant's *reason* had subjected all proofs of God's being to analysis and proved that they rested on a logical error.

We remember that *The Only Possible Proof of God's Being* was the first work which brought its author a measure of literary fame. The young philosopher was then of the opinion that such a proof could rest on a foundation such as the ontological argument. The argumentation runs as follows: we think of God as the most perfect being; if the predicate of existence cannot be applied to this being, then it is not perfect, and we fall into a contradiction. Now, Kant laughs at such methods of confirmation. He says to the theologians: you contradict yourselves as soon as you introduce "reality" into the understanding of a thing which you are only prepared to think possible. A real thing (*Gegenstand*) has no more predicates than a possible thing. One hundred real talers are not worth more than one hundred possible talers. The difference lies simply in the fact that I have the real talers in my pocket. The concept (*Begriff*) is not yet existence (*Sein*).

In the second proof for God's existence — the cosmological — Kant discovers a similar error. The existence of the world must be preceded by the prime cause, and that must be God. In Kant's view such postulate is acceptable on condition that man not take this thought to be matched by a real state of affairs. Again, the concept is not yet existence.

In the third method of proof — the physico-theological — the issue is a general teleology which we discern in nature. Does it not bespeak the wisdom of the creator? That is possible, Kant answers, but in this instance God is not the creator of the world but merely a builder who uses the finished material ready to hand. But there is something else which is more important: an error is being repeated, namely, an arbitrary notion of the causal dependence of the world is being posited as real.

Therefore, Kant sums up, the cosmological method of proof lays the foundation for the physico-theological method, and the ontological method lays the foundation for the cosmological approach, and since there are only these three proofs, the ontological argument is the only one possible. Has Professor Kant repeated the conclusion of his work as *magister*? Yes, but with one essential limitation: "And since besides these three there is no other path open to speculative reason, the ontological proof from pure concepts of reason is the only possible one, if indeed any proof of a proposition so far exalted above all empirical employment of the understanding is possible at all."[102]

Hegel was not thrown by either Kant's irony or his logic. Proofs of the existence of God are then free of error, he maintained, when they are conducted with the help of dialectical logic. Kant (he was then long gone) could have objected: dialectic is a logic of appearances. "Now I maintain that all attempts to employ reason in theology in any merely speculative manner are altogether fruitless and by their very nature null and void, and that the principles of its employment in the study of nature do not lead to any theology whatsoever. Consequently, the only theology of reason which is possible is that which is based upon moral laws or seeks guidance from them."[103]

Let us sum up: Kant denied the Lord God the throne of nature and logic, but left him an unrestricted domain: morality. Kant did not need God to explain the phenomena of nature; but when it came to man's conduct he did not say categorically that without God there was no way; to explain things here the idea of a higher being might be of some use.

That provoked Heine's wicked jest: "You mean we can go home now? Just a moment, we are not done yet! Tragedy is followed by farce. Until now Immanuel Kant has played the merciless philosopher, he has stormed the heavens, he has routed its whole garrison, the supreme ruler lies unproved, in his own blood, now there is no longer any all-embracing mercy, no fatherly love, no reward in the other world for the restraint shown in this one, the immortality of soul is breathing its last — there is groaning and moaning — and the old Lampe stands there with his umbrella under his arm, a mournful observer, cold sweat and tears running down his cheeks. Then Immanuel Kant takes pity and shows that he is not only a great philosopher but also a good man, and he ponders, and then, half in earnest and half in jest, he says: 'the old Lampe must have a God, otherwise the poor man will never be happy — man should be happy on this earth — so practical reason tells us — oh, well — let practical reason guarantee the existence of God'. In consequence of this argument Kant draws the distinction between theoretical reason and practical reason, and with the latter, as with a magic wand, he resurrects the corpse of deism which theoretical reason had put to death."[104]

This is a brilliant passage; it is far more telling about Kant than the accusation that he was bourgeois. We can have no knowledge about God, Kant says, only faith remains. But what is faith? One of the last sections of the *Critique of Pure Reason* is entitled "On Opinion, Knowledge and Faith." Faith can make less of a claim to truth than knowledge (although it is more reliable than mere opinion). If the truth of judgment is only subjectively valid, its objective ground being insufficient, then it is faith. Knowledge represents both the subjective and the objective acceptance of the truth of judgment.

"Thus I had to cancel (*aufheben*) knowledge in order to make room for faith,"[105] the thinker audaciously declared in the preface to the second edition of the *Critique of Pure Reason* — a book that makes the highest claims on knowledge. The audacity must be derived from the ambiguity of the proposition. And it is here that Kant shows his true face. He uses the verb *aufheben* which literally means "to elevate" (*sublevare*), then "to do away with" (*beseitigen; tollere*), and, finally, "to maintain," "to preserve" (*verwahren, bewahren; conservare*). Kant removed knowledge from fields which did not genuinely belong to it, he elevated it to a higher standpoint, made it subject to the rules and laws of the critique of pure reason and thereby preserved its purity and power.[106]

That is the fate of knowledge. The last word on faith has not been said yet. There are three kinds of faith. Faith which a man has only in one instance or another Kant calls pragmatic. Such faith is worth no more than a ducat. "Frequently man expresses his convictions with such confidence and unyielding fervour that it appears he does not doubt their correctness. A wager will stop him short. Sometimes it seems that he has a ducat's worth of conviction, but not enough to go up to ten ducats. For he takes up the wager of one ducat, yet when things come to ten ducats he begins to discover that there are things he did not notice before, namely that he could be in error."[107]

A theoretical faith Kant calls doctrinalism. The believer is ready to gamble all his possessions on the notion that at least one of the planets in our system is inhabited. That is an example of doctrinal faith. Kant sees the doctrine of the existence of God as belonging here. Doctrinal faith has a weak point: often the complexities entailed by speculation push us away from it, although we steadfastly return to it.

With moral faith things are entirely different: there the problem of whether a judgment is true or not does not arise at all: "Since the moral precept is my maxim (as reason commands it to be) I inevitably believe in God's existence and in a life after death, and I am certain that nothing can shake that faith because that would destroy my moral principles themselves, which I cannot deny without becoming abhorrent in my own eyes."[108] Faith in God in this instance does not mean that man must reflect upon his existence, but only that he must be morally good.

Thus Kant's thesis that knowledge stands above faith supplies its own stipulating antithesis: it does not allow its becoming entangled with knowledge because it is realized exclusively through the activity of practical reason.

The critique of pure reason concludes with a program for the future. Kant does not yet have any new "critiques" in mind. The critical labour has for the time being been brought to a conclusion. The field has been weeded, ploughed, fertilized: now it must be planted. Kant knows what is going to sprout: the positive principles of metaphysics.

Hegel will bring the term metaphysics into disrepute; by it he means petrified thought drawing false speculative conclusions from the premises of formal logic; true to life philosophy Hegel calls dialectic. For Kant it is the other way about: dialectic is an illusory logic, metaphysics is the wisdom of the world. He speaks of metaphysics with great respect. For him it is the "perfection of all the culture of human reason;" he who becomes disappointed in it will sooner or later return to it as if "to a lover after a quarrel."[109] It is inconceivable that man's spirit could ever turn completely from metaphysical speculation: no sooner would man stop breathing because of the fear there might be pollution in the air.

The problem is that it is possible to "blunder about any which way in metaphysics without any fear of being trapped in a falsehood."[110] There are no tests of veracity in metaphysics of the kind that the natural sciences have at their disposal. That is why metaphysics has not been a science up to now. But it can become a science. In relation to other sciences it has an undeniable advantage: it can be brought to a state of completion because there are no new discoveries possible for it which are unavoidable in the other sciences; its source of knowledge is not the multitude of things in the external world, but reason itself; and when reason has clearly and completely laid out the fundamental laws of its own capacity (*Vermögen*) there is nothing left for it to know.

Kant predicts a rebirth of metaphysics in accordance with an as yet unknown plan. In the last pages of the *Critique of Pure Reason* the reader finds a plan for the rebirth of philosophy. It is not at all original. Kant divides the whole system of metaphysics in four parts — ontology, physiology, cosmology and theology. Ontology is the science of the universal principles of being; physiology — as Kant understands it — is the science of nature, further divided into physics and psychology; cosmology is the science of the world as a whole; theology treats of God. Wolffian metaphysics was constructed in the same manner. Kant actually gives no answer to the question put at the beginning of the critique: "how is metaphysics possible as a science?" He has destroyed all the dogmatic constructs in this sphere with his transcendental dialectic but does not himself get further than the declaration that a new scientific philosophy is necessary.

The publication of the *Critique of Pure Reason* caused no sensation. Nor did Kant expect it to produce a stir, but what did happen exceeded his worst expectations: the book was not noticed at all. It sold so badly that Hartknoch considered selling the rest of the stock as bulk paper. (Today a copy of the first edition brings in up to 7000 DM in German rare book shops.)

The book was difficult reading and fairly uninteresting. No one came to Kant's defence, no one attacked him; now and then he would hear complaints about the book's unintelligibility. Moses Mendelssohn read a few pages and put the book down. Hamann read galleys with the author but appears to have had no great joy in doing so. Such a pedestrian book, he wrote to Herder in May 1781, was unworthy of its author and of the concept of pure reason, which was opposed to his own, lazy reason. He called Kant a "Prussian Hume." Herder opined: "Kant's critique is a tough nut to crack, doomed to remain unread . . . I do not know why all this heavy gossamer . . ."[111]

As soon as the book was out Kant sent complimentary copies to his friends and renowned scholars. The first on the list were Minister von Zedlitz, Moses Herz, Mendelssohn. In August 1781 Kant remembered the court chaplain Johann Schultz who had written a review of his last dissertation and praised him as the best philosophical mind of his time. He was also sent a copy; it took him two years to reply.

The letters Kant received spoke of everything under the sun, except the *Critique*. Hartknoch reminded Kant that he was expecting new works from him; he offered to use the books he still had in stock and as a sign of friendship sent the philosopher a pound of tea. Berens the merchant begged Kant to lead his prodigal son back to the paths of truth. The mathematician Bernoulli was interested in Kant's correspondences with the late Lambert; the philosopher informed Bernoulli that his book had come out and gave him a precis of its main ideas, but Bernoulli was interested only in Lambert.

From Petersburg came the news of a mysterious epidemic: it had come from Siberia and spread throughout European Russia. That was an answer to Kant's query about the source of a disease which had spread through Eastern Prussia in the spring of 1782. The disease was not lethal, but patients felt very ill for days. They suffered with fever, vertigo, coughing, and nasal congestion. It struck largely those who were often out and about and moved in crowds. Diaphoretics and blood-letting were the usual methods of therapy; Spanish flies were administered to ease the coughing. The peculiar disease was called influenza. Kant remembered that seven years before there had been an outbreak of influenza in London. This led him to publish a notice, "Advice to the Physician" in the local paper, to which he appended a translation of an English physician's report about the London epidemic of 1775. Kant himself did not fall ill: he was saved by the application of his own system of hygienic rules. He was interested in the spread of the epidemic. Having considered the increase of contacts between peoples brought about by trade, he predicted that new epidemic outbreaks would sweep Europe.

The first personal response to the *Critique* (judging by the preserved correspondence) came from Kant's brother. They rarely wrote to each other. After the university Johann Heinrich settled in Kurland. He made a love match. "I am happier than you: follow my example," he had written his brother in 1775. Now, in September 1782, Johann Heinrich informed his brother that several years before he had taken up the post of a preacher somewhere between Mitau and Riga. He thanked him for the worthy book — it was going to help him in the running of his affairs. The book came in handy in his new surroundings: the soil was rich, the garden was marvelous, the children were growing up and he had no cause for complaint. Only the social environment left something to be desired, his flock was common, there was not a single nobleman in it. Johann Heinrich wanted to know about his relatives — his sisters, his uncles and aunts, what was his brother doing? The letter closed with the following passage: "All thinkers hereabouts are talking about the critique of purified reason. I hope that you will not stop now. Could your brother not make a small request, namely that he should know about your work a little before the rest of the world?"[112] Then there was a postscript from his sister-in-law: she was enthralled by the book on housekeeping (*Haushaltsbuch*) and thanked "her brother, sir" from the bottom of her heart.

In the meantime two anonymous reviews had appeared. The first was published in the January issue of *The Göttinger Learned Messenger.* The reviewer ascribed to Kant views that were not in the *Critique* and said nothing about what should have been clear from the first line. Kant could not recognise his own thoughts. The principal idea of the critique was reduced to the following: all our knowledge comes from certain modifications of our selves, which we call perceptions (*Empfindungen*). Where they were to be found, whence they came, is absolutely unknown to us. The review accused Kant of Berkleyanism.

The author of the review was Christian Garve, a representative of "popular philosophy" and a well-known translator of English and Greek, who had had to give up a university career due to his ill health. When the *Critique* came on the market he happened to be passing through Göttingen and promised to analyse the book even before he had seen it. As soon as he had read it he realised that he could not do much with it. But it was too late to retreat. So the review was written with much anguish; it finally came into the hands of Professor Feder, a member of the editorial board who without consulting Garve shortened the text and put in a few of his own contributions—the comparison between Kant and Berkeley, for instance, was his. Garve renounced the authorship of the review. When Kant publicly asked that the Göttingen reviewer reveal his true name, Garve wrote an agitated letter to Kant, explaining the state of affairs, reproaching himself for having accepted a task which he was not fit to discharge, repenting the deed but placing the main responsibility on the member of the editorial board who had left barely a tenth of his text in the published review. (That was not quite true: Garve's review was later printed in entirety in *The Allgemeine Deutsche Bibliothek*, and Kant could see that there was very little difference between it and the Göttingen variant; he was dissatisfied and said that they must have thought him an idiot.) Kant wrote a kind and copious answer to the penitent letter, suggested that the acquaintanceship be maintained, calmed the hapless reviewer and said that his only concern was a peaceful discussion of the issue at hand. He did not like bitter scientific squabbles but, unfortunately, nothing else could be expected. "Oh, you weak men who claim that your only concern is for truth and the dissemination of knowledge, in fact your major concern is vanity!"[113]

Another review appeared in Gotha, in August 1782. Its thoroughly mediocre author, a certain Ewald, could not evaluate that text at all. Fortunately he limited himself to merely repeating the introduction and several sections of the book.

This was all very depressing, the more so as Kant did not think that a great philosopher was doomed to be incomprehensible. On the contrary, he firmly believed that even the most difficult work can and should be made accessible to all. Yes, here was a unique criterium of truth: the test of popularity. He expresses this in the letter to Garve as follows: "You take pleasure in mentioning its lack of accessibility as a proper reproach that one can direct at my work; and in truth, it must be possible to popularize every philosophical work, otherwise nonsense can hide under the veil of seemingly brilliant and profound thought. However, popularity cannot find its origins in research which attempts so much and reaches so far."[114]

The major work now behind him, Kant could worry about it being correctly understood. While he was still correcting galleys of the *Critique*, Kant told Herz in a letter that he wanted to make the content of his work accessible to all. After the cool reception of the book Kant was strengthened in this intention. But his work was extended: Kant began his work on the metaphysics of morals at this same time. Besides this, he was waiting for further reviews. They appeared; Kant no longer delayed with the completion of the compendium to the *Critique*. In the spring of 1783 the book came out, entitled *Prolegomena for a Future Metaphysics Which Can Become a Science.*

Prolegomena means "introductory remarks." The exegesis of the critique was significantly shorter than the work itself, but just as impenetrable. In the foreword the author wrote that the prolegomena were not meant for the students but for future teachers. The stress was placed on the problem of metaphysics. Stung to the quick by the charge that he was trying to do away with philosophy. Kant now strongly emphasized his devotion to the cause of speculative philosophy: if it only accepted his foundations the future of philosophy was rosy. (How it was to transform itself into science remained as vague as before!) At the end of the work Kant took up the published reviews. This was the place in which he invited the Göttingen reviewer to join him in a public debate. As for subjective idealism, Kant compared his theses with Berkeley's and assured the reader that his views were dramatically opposed to those of the contemporary idealism.[115]

The accusation of Berkelyanism continued to rankle. In order to lay it to rest Kant wrote another chapter for the second edition of the *Critique of Pure Reason*, entitled "The Refutation of Idealism;" in the preface, too, he said that it was necessary to prove the real existence of the things external to us. It was a scandal of philosophy "to have to accept the existence of the things around us on faith." Prompted by such considerations he significantly compressed the material dealing with the generative role of productive imagination. The cuts here are so important that the present day researcher must turn to the text of the first edition in order to obtain the complete picture. The text of the following three editions, which came out in Kant's lifetime, remained unchanged.

The second edition came out in 1787. The ice had been broken: critical philosophy was gradually making its way among the readers. The problem of popular reception, once a source of anxiety for the author, resolved itself in a most unexpected fashion. Johann Schultz, who had received a complimentary copy in August 1781, now spoke out. He had studied the book for two years. (Mendelssohn needed more time; two years later he wrote to Kant: "Your critique of pure reason has become my criterion of health. As soon as I manage to convince myself that my strength is returning I turn to this work, which saps my nervous energy; I have not yet lost the hope that I shall live long enough to be able to think it through thoroughly.") Schultz had studied Kant's work with care and written a detailed review. He sent it to Kant before it was published: Kant, he thought, was "the best interpreter of his words," and the public were interested, above all, in an adequate transmission of the sense contained in the reviewed book.

Kant was delighted. Not only because he agreed with the review but because he had finally found a man who understood him correctly. Kant made a number of additions and recommended to Schultz to publish the review as a separate booklet. Schultz took his advice: he claimed the right to popularize Kant's thought because the prolegomena could not accomplish this task; by themselves, without a previous reading of the critique, they were quite incomprehensible. The review turned into a book: *An Elucidation of Professor Kant's Critique of Pure Reason* is a conscientious commentary on Kant's epistemology.

On 11 February 1874 *The Gotha Learned Gazette* published a short notice stating that the Königsberg court chaplain Schultz was working on a popular exposition of Kant's *Critique of Pure Reason*. The notice said: "One of Professor Kant's favourite ideas is that the final goal of the human race is the accomplishment of a perfect constitution, and he wishes that a philosophically inclined historian would take it upon himself to construct a history of mankind according to this plan and to show how near, or how far, man in different times comes to that final goal, and what there is left for him to do in order to reach it."[116] It was not a chance notice. It speaks about Kant's new interests, new plans.

Two things fill the soul with ever new and growing wonder and awe the more we think of them — the starry sky above me and the moral law in me.
— KANT

CHAPTER FOUR

The Idea of Personality

In the spring of 1784 Kant celebrated his sixtieth birthday. He entered upon the day in the full powers of his intellect (although the same could not be said of his body). Old age was still far away. The most productive period of his life was about to begin. Three quarters of his life were behind him, but barely half of his works had been written. Now they were tumbling out one after another — books, essays, treatises, reviews. Kant widened the frame of critical philosophy; the principles of epistemology were being applied in other areas. The established truth was tested time and again; if it withstood this, it ws more firmly established — if it did not, it was replaced by a new truth, which was in turn tested, made more precise and more complete. "Work is the best way to enjoy life," said the seventy-five year old Kant.[117] When he was sixty he was already of that opinion.

The official celebration took place early, on 4 March, the last day of the semester. The students gave their professor a commemorative medal. On the obverse was the likeness of the celebrant, on the reverse an allegorical representation: the leaning tower of Pisa with a plummet hanging from its top and a sphynx at its feet. The idea was Mendelssohn's. We know with how much difficulty he had read the *Critique of Pure Reason*. The image of the tower which had been constructed by the philosopher came from the *Prolegomena*. To Mendelssohn Kant's philosophy looked like an edifice ready to topple, so he suggested a verse to the effect that although the tower seems ready to collapse it does not. The allusion was too transparent, and a different text was chosen: *Perscrutatis fundamentis stabilitur veritas* (Truth can only be ascertained by a thoroughgoing re-examination of its foundations). Now it was only the sphynx that indicated the unusual nature of Kant's ideas. The year of the philosopher's birth was mistakenly given as 1723. (The year of his birth was often pushed back a year; probably because he was in the habit of speaking not of the year of his birth but of the number of years he had completed.) His name was misspelled as well. And the likeness was not very good . . . In a word — the celebrant did not like the medal, especially the allegory on the reverse.

In the prolegomena the issue was precisely that human reason periodically erected towers and then pulled them down to inspect their foundations. Naturally, Kant did not think that was true in his case: he began with the foundations and only after having convinced himself that they were adequate built on — upwards and outwards.

In relation to the expansion of the construction we can note the above-mentioned ideas about the philosophy of history. The Italian Vico christened the philosophy of history "una scienza nova" when he first thought about the ways of considering the development of human society. If epistemology had a history of a thousand years, the theory of the historical process came into being before Kant's eyes.

Fruitful impulses came from abroad. In 1750 the future minister of Louis XVI, Turgot, had already delivered his famous lecture at the Sorbonne about the development of human reason. Ambition, love of fame and vanity, he said, are the causes of incessant changes on the stage of the world; they saturate the earth with human blood. Yet even in the process of devastating changes they have caused morality and reason to assert themselves; isolated nations are drawing closer together, trade and politics bind all the parts of the globe together. And the whole mass of the human kind that has until now in turn known peace and distress, years of happiness and years of want — now steadily advances, although with small steps, towards a greater perfection.

Many at that time shared Turgot's thoughts. Voltaire was already working on his *Essai sur les moeurs et l'ésprit des nations*, the basis of which was the idea of the progress of knowledge and enlightenment. In Germany Isaak Iselin's book *On the History of Mankind* was published, in which world history is divided along the lines of a tripartite schema. The first level is characterized by the predominance of feeling, it is the condition of a primeval "simplicity," of the childhood of the human race; the peoples of the east have remained on this level. The Greeks and the Romans reached the second level, where imagination superseded the strength of feeling; virtue and enlightened wisdom made morals milder, although barbarism was not yet completely surpassed. It was only after a thousand year reign of darkness (as Iselin called the Middle Ages) that the European peoples reached the stage of civilization where reason ruled over both the feeling and the imagination. The history of mankind is thus a gradual and steady progression of reason and morality towards ever greater perfection.

Two more names and works must be mentioned so that the backdrop against which Kant and his pupil Herder stood might be complete: Lessing's *The Education of the Human Race* (1780), and Adelung's *A Cultural History of Mankind* (1782). Lessing patterned his schema of development upon the history of religion: Christianity was considered only as a preparatory stage in the evolution of mankind towards the ideal of humanity. Adelung discussed the problems of cultural history: the determining factor here was the increase in population.

The first part of Herder's *Ideas of a Philosophy of the History of Mankind* came out in May 1784, Kant's *Idea for a Universal History from a Cosmopolitan Point of View* in November of the same year; the latter, however, appears to have been written in the spring. The above-mentioned notice in *The Gotha Learned Gazette* announced the forthcoming work: in the beginning of the treatise Kant mentions the notice and remarks that he had had to develop his views with more precision than he had done in conversation with a visiting scholar; they were now ready for publication. Kant's *Idea* developed independently of Herder's *Ideas*.

It is necessary to remember this, since the pupil accused his teacher of plagiarism (although not publicly). We shall show how matters came to an open conflict; first, however, we shall acquaint the reader with the contents of the essay, which sets out the fundamentals of

Kant's philosophy of history in a lapidary form, in a number of theses. The essay, published in *The Berlin Monthly*, was very well received. It was reprinted a number of times during the philosopher's life. Schiller's familiarity with Kant's philosophy, in which he quickly became very adept, began with this essay.

The *Idea for a Universal History* begins with the investigation of a problem which was more or less generally posed in the eighteenth century: *viz.*, does social life follow certain natural laws? It might appear that these are more subject to chance in the life of a man than the contraction of marriage. Yet, the accumulated data indicate that this process is subject to laws in large countries, just like the fickle weather whose vagaries cannot be predicted in specific instances but which usually sustains the growth of plants, the flow of rivers, and other natural structures steadily and without interruption. Individuals and even entire nations do not much consider the imperceptible fact that they follow a natural purpose unknown to them, as they would follow on a lead, and that they jointly strive to achieve this goal in the pursuit of their aims — each in his own way, sometimes irrationally and to the detriment of others. Vico had already made the claim that individual aims stand in a sharp contrast to the results of social activity. Herder restated the observation, and Hegel would call it the "cunning of reason."

Kant does not believe that the individual exhibits the working of a rational purpose; stupidity, childish vanity, wickedness and destructive passions are more often the actual motives of human behaviour; but if we stand away from them then the general progress of history reveals a purpose common to all mankind. In this sense man's natural abilities constituted so that they require the application of his reason, develop fully not in the individual but only in the species. The individual is mortal, the species is immortal. It takes countless generations, each passing its enlightenment on to the next, to bring the species' capacities to their full development.

What means does nature use to develop the capacities planted in men? The cause of the regulative order in mankind is a primeval antagonism which Kant calls "asocial sociability," *viz.*, the inclination to come together in society and at the same time to oppose this society, threatening its unity with disintegration. Driven by ambition, the thirst for power or greed, man carves out a place for himself among his fellow men, whom he cannot stand, but without whom he cannot prosper.

These are the first steps from barbarism to culture. Under the conditions of an Arcadian pastoral life, with perfect concord, justice and mutual affection, man's talents could not manifest themselves, and man, as tame as a sheep, would pursue an existence no more worthy than that of domestic animals. Therefore, blessed be nature for the quarrelsome disposition, for the enviously competitive ambition, for the unquenchable thirst to subjugate and rule! Man wants concord, but nature knows better what is good for his species and leads him to the path of discord.

Where does this path lead? Kant is an optimist; he is convinced that in the final analysis the road leads to the constitution of a just civil society in which the individual achieves the greatest freedom which does not abridge, and is not abridged by, the greatest freedom of all. Antagonism will exist in such a society, but it will be tempered by laws. Only such conditions favour the most complete development of the potential implanted in human nature.

The constitution of a universal just society is the most difficult task, and it is the last task which the human species will accomplish. Man by his animal nature wants to be ruled; as a rational being he creates laws to limit individual caprice in all, but his selfish animal inclinations move him to make an exception of himself. Every man endowed with power will abuse his freedom if there is no one above him to curb his excesses by the force of law. Therein lies the difficulty of the task set before man. It is impossible to solve it in entirety; but nature compels us to strive for a solution. Three interrelated conditions are absolutely indispensable to the solution: a correct conception of the constitution of the state, centuries of experience, and good will. Kant harbours no illusions regarding the time when all this will come to pass: not so soon, but not so far in the future, either — after many fruitless attempts.

The creation of a perfect civil society within a state depends upon yet another condition: the development of laws governing relations among states. This is merely the extension of the principle that binds individuals in the state and prevents them from exterminating one another.

> Nature has thus utilised again the intolerance of men, even of large societies and political bodies, as a means for establishing a condition of peace and security within the inevitable antagonisms between them. Through war, through the taxing and ever-

> lasting accumulation of armament, through the want which any state, even in peacetime, must suffer internally, nature forces men to make at first inadequate and tentative attempts; finally, after devastations, revolutions, and even complete exhaustion, she brings them to that which reason could have told them at the beginning and with far less sad experience, to wit, to leave the lawless condition of savages and enter into a league of nations. In a league of nations even the smallest state could expect security and justice, not from its own power or by its own decrees, but only from this great league of nations (*Foedus Amphictyonum*), from a united power acting according to decisions reached under the laws of their united will.[118]

Kant was quite immersed in the problems of world history when he received a letter from Jena, containing an attractive proposition. The well-known philologist, Professor Schütz, fulsomely praised the *Critique of Pure Reason*, made a few of his own remarks and then announced that next year he was planning to publish a *Universal Literary Gazette* which would be dedicated exclusively to reviews of newly published books. Kant's collaboration would greatly add to the prestige of the new publication. Could he not take it upon himself to review Herder's *Ideas for a Philosophy of the History of Mankind?* Kant answered that he would.

We left Herder in Bückeburg, ministering to his flock and contributing to the literary movement "Storm and Stress." In 1776 he was called to Weimar, to be head of the Lutheran church there. The spirit of freethinking reigned at the court of the liberal duke Karl August von Sachsen-Weimar. Herder became an enthusiast of Spinoza and began to study biology with Goethe. In 1784 Goethe made an important discovery: he found in man the intermaxillary bone, which animals do not possess, and his discovery was considered an argument in favour of a fundamental difference between man and animal; the discovery was not made public, Kant knew nothing about it, but Herder, naturally, did. We therefore have every reason for seriously considering the propositions regarding evolutionary development contained in the pages of Herder's major philosophical work *Ideas for a Philosophy of the History of Mankind.*

Herder prefaces the analysis of the development of the human species by a survey of the history of our planet. In his description of the origin of the earth and its place in the cosmos Herder avails himself of his teacher Kants's *Universal Natural History and Theory of the Heavens*. He also discusses the geological history of the earth, its flora and fauna. In extremely vague allusions Herder speaks about the natural beginning of life, which occurred, in his opinion, in the water. His argument about the continuous development of animal species is equally vague. The scientific conjectures in *Ideas* are interwoven with views on the transmigration of souls and their immortality. The goal of evolution lies in the infinite; Herder wonders what the destination of man's evolution toward perfection is. On earth man is his own limitation; Herder's imagination sees further progress in life in the realm of the extraterrestrial, the world to come. Thus ends the first part of his book, the part Kant received. The history of mankind was actually not the subject at all.

But that was precisely what the reviewer found interesting. Kant saw nothing in Herder's constructs except flippant audacity. The pupil was swift in solving problems that his teacher had in his youth left unsolved: in the *Universal Natural History and Theory of the Heavens*. Kant doubted that it was possible to apply the principle of evolution to organic matter. Herder's style also made an unfavorable impression on Kant: it was emotional, occasionally bombastic, bereft of clarity and the power of proof. Instead of logical precision and strength the reader finds only vague, ambiguous illusions. Kant selected several quotations which showed how Herder judged things about which people had previously only fantasized. The notions concerning organic evolution appeared to Kant so incredible that they "clearly went beyond all human reason." (We must not judge Kant too severely here. Even Hegel, whose philosophy was thoroughly permeated with thoughts concerning evolution, could not bring himself to apply these ideas to the world of nature.)

Kant becomes ironic when Herder wants to ground the immortality of soul in the general theory of evolution. Even if there are beings possible whose organization is superior to man, there is no reason to conclude that one and the same individual will reach that higher state. There are various beings which evolve to higher levels of organization. The caterpillar evolves into a butterfly, but between the two stages there is no death, only the stage of chrysalis. In conclusion of his review Kant expressed the wish that the pupil "constrain his lively genius. It is to be hoped that philosophy, whose concern consists more in the pruning than in the sprouting of superfluous growth, may guide him to the consummation of his enterprise, not with hints, but precise concepts, not through supposed, but through observed laws, not through the intervention of flighty imagination, whether metaphysical or sentimental, but rather through the exercise of careful reason in the execution of his bold project."[119]

The review appeared in one of the first issues of the *Jenaische Allgemeine Literaturzeitung*; it was unsigned but Herder immediately recognised the argumentative handwriting of his teacher. Kant's name was printed among the names of contributors to the new paper, and so the identity of the reviewer was a public secret.

Herder was offended, embarrassed, and desirous of striking a blow in return. In a letter to Hamann he railed against the *Critique of Pure Reason*: "I shall be mightily pleased when I have crushed or annihilated his idol of reason."[120] But these were distant plans. In a letter to Jacobi he gave his views on Kant's *Idea for a Universal History from a Cosmopolitan Point of View*. The head of the Protestant church in Weimar risked words in the letter to Jacobi which would be inimical to censorship: "I wish the heavens would inspire you to write a few sentences about the absurdity of this *Idea* (for the rest and the whole plan were purloined from *Idea*). For example: man is an animal that pines for a master; man does not exist in himself but for his species; he develops all his powers within the species, and in the final analysis everything is constructed for the purpose of political antagonism and a most perfect monarchy, nay, a coexistence of many most perfect monarchies, ruled over by pure reason *in corpore* etc., etc."[121]

In the blindness of his anger Herder was as wrong in his accusations as Kant had been in his review. Each of them was not only unwilling to see anything positive in the other, but willfully distorted and simplified his opponent's thoughts in the exposition of them. Kant had borne an analogical situation with dignity: the Göttingen review had not damaged his personal relationship with Garve; the duty of the author of the *Critique of Pure Reason* was to make the ideas contained in the book understandable to the reader, to help the reader understand him; the job of the reviewer was only to provide arguments. The vehement Herder wanted to discredit his opponent at any price. The minister was boiling with indignation. The polemic with Kant quickly turned into an open quarrel, although the latter tried very hard to maintain his attitude of irony.

Kant's review appeared just as Herder was about to complete the second part of *Ideas*. He hastily added several attacks on the critical philosophy. But before Herder's book saw the light of day Kant had to read a sharp answer in *The Teutscher Merkur*. Many were unsatisfied with Kant's review (which further fueled Herder's ire). In a letter to the aggrieved Herder the philosopher and poet Knebel called Kant "a fool who measures wisdom with his own yardstick", and expressed the concern that this "learned ass" might slow down Herder's progress.[122] Wieland, the publisher of *The Teutscher Merkur*, was also dissatisfied. In his journal an answer to Kant appeared signed by "A minister from ***."

The anti-reviewer accused the reviewer of "metaphysical routines" which prevented him from seeing the living thoughts in *Ideas*, which enquired into new experiential data. The lengthy discussion contained a number of accusations and sharp words; the most piquant bit was the fact that under the mask of the minister there hid Wieland's son-in-law, the budding philosopher K.L. Reinhold, who had not yet read Kant's major work. (Later, when he read the *Critique*, he became enthralled by it and spread its thought far and wide.)

Kant did not restrain himself; he had to express his opinion of the anti-reviewer publicly. His opponent, he wrote in the *Jenaische Allgemeine Literaturzeitung*, accused him of being a metaphysician who did not acknowledge empirical data and was ossified in fruitless abstractions; on the contrary, he had in his review relied on the empirical facts of anthropology and other sciences. Kant was particularly incensed by the "minister's" remark that reason could not be allowed to retreat in face of any bold idea; in the answer Kant again insisted that Herder's ideas were unscientific in nature.

The author did not think it was necessary to defend himself. His answer was not a response but a new attack. Herder did not mention Kant by name once in the second part of *Ideas*, but between lines the conflict widened. For instance, he attacked Kant's fundamental proposition of *Idea for a Universal History* that conflict defined the original condition of society. He ultimately — incorrectly — quotes Kant's essay: "An easier but more repugnant principle of the philosophy of human history would be: man is an animal who needs a master or who from that master or his relation to that master expects the happiness of his ultimate purpose." There was nothing about "happiness" in Kant; Kant only wanted to state that men abuse their freedom and therefore were in need of a "master": as such he introuduces the human species, which finds the institutionalized form for its existence in the state. Herder's angry philippics against the state were directed at Kant as well, who saw this institution as inevitable and tried only to reflect on how it could be perfected. For Herder the state was a machine which had to be destroyed every now and then. So he altered Kant's aphorism: "man is an animal who needs a master" was changed to: "the man who needs a master is an animal; as soon as he becomes man, he will no longer need a master."[123]

Yet another arrow was directed at Kant: it was the again completely groundless charge that critical philosophy scorned the human individual in so far as it saw the species as the self-sufficient form of being.

> When someone says that not the individual but the species is educated I find it incomprehensible since the species is only a universal concept in so far as it is embodied in particular beings. If I would grant this universal concept all the perfections of humanity, culture, and the highest level of enlightenment, which is what an ideal concept implies, I would have said as much about the true history of our species as if I had spoken about animality, minerality or metality in general and conferred upon them the most magnificent predicates, which however contradict one another in single individuals.[124]

When the second part of *Ideas* came out Kant received the book from Hamann and studied it carefully. Hamann informed Herder that Kant had spent more than a week on it, which was unusual. A new review appeared in the *Universal Literary Gazette* — this time benevolent right at the beginning: Kant praised the intelligent handling of ethnographic sources and their masterful interpretation which he said were accompanied by excellent judgments. But he becomes ironic when he deals with the metaphoric manner of interpretation, in which, in Kant's view, synonyms replace arguments and allegory replaces the truth. He gave examples of how Herder contradicted himself.

Kant, of course, did not miss the attacks on him scattered through the book: Herder emphasizes above all the happiness of the individual, which he opposes to the interests of the state. Men can be happy in various ways, replies Kant. The notion of happiness that each man forms for himself was not the heart of the matter for Kant; rather only the "endlessly growing and progressing activity and culture, the apex of which can only be the product of a state constitution ordered according to the concepts of human rights"[125] — this is the true aim of providence and enlightenment. If the happy islands of Tahiti, which civilization has passed by are the ideal, the question arises of whether man is at all necessary for such happiness, and whether he could not be replaced by happy sheep and cattle. Herder termed Kant's principle "light, but repugnant." Man can easily accept such a principle, as it is confirmed by the experience of all times and peoples. But why should it be "repugnant" — could it have meant that a repugnant man had uttered it?

A smile also permeates Kant's critical remarks on Herder's statement that there were no attributes of the species which were not also present in the individual. Of course it is nonsense to say that no horse has horns but that the horse species is horned. But the human species as a whole does have certain characteristics which are absent in the particular individual. Only the species actualizes its destination, that is to say only the species is involved in the activity of development and is capable of fully achieving its peak.

With this Kant concluded the polemic. He firmly refused to review the third part of Herder's *Ideas*. One can hear echoes of the polemic in some later works, for instance in the treatise *Conjectural Beginning of Human History* (1786) where he — very carefully — heaped irony on the old Old Testament, so dear to Herder's heart as a minister. But these are only echoes. For Kant the issue had been laid to rest.

Herder, however, had only just begun. Offended to the innermost reaches of his soul, he had not even completed his major work when he began squandering countless hours and a great deal of energy in vain attempts to undermine the critical philosophy. To the *Critique of Pure Reason* he counterposed a *Metacritique of the Critique of Pure Reason*, to the *Critique of Judgment* his *Kalligone*, to Kant's *Perpetual Peace* he offered his own work with the same title. Kant paid no attention to attacks.

His attention was drawn to another polemic, which was then raging among the German intellectuals. That was the famous "pantheism conflict." Four years after Lessing's death (he died just before the appearance of the *Critique of Pure Reason*, in 1781) it became known that the great literary critic had become a Spinozist. The philosopher Jacobi broadcast the news to the world. Having found out that Mendelssohn was planning to write an essay about Lessing he informed him of the conversation he had had with the deceased. The conversation had been about Goethe's *Prometheus*, a poem about which Lessing had expressed his complete agreement; orthodox notions about God no longer mattered to him in the least, God and the world were one and the same. The startled Jacobi had not expected this and he asked Lessing in confusion whether he was not, then, in substantial agreement with Spinoza. Lessing had answered that he had no one else in mind and wanted to know whether Jacobi knew of a better man. Spinozism was in the Germany of the day a synonym for atheism. Men could still remember the burning of books and the persecution of those who adhered to the Dutch pantheist's system of thought.

Mendelssohn did not believe Jacobi's story; a correpondence between them began which the latter published in 1785, after Mendelssohn had published his book about Spinoza and Lessing (*Morning Hours*). Mendelssohn answered with *Letters to Lessing's Friends*, which was followed by Jacobi's *Against Mendelssohn's Accusation Concerning the Letters on Spinoza's Doctrine*.

Jacobi attacked Spinozism; he saw it as a thoroughgoing expression of a rationalist philosophy which was incompatible, in his opinion, with either faith or true knowledge. Reason was for him the capacity for an unmediated apprehension of God's being. Mendelssohn did not share Spinoza's views either; in his polemic with Jacobi he did not defend pantheism, but Wolff's rationalism. His major concern was a defence of the late Lessing from the charges of an odious godlessness.

Both men asked Kant to come to their aid. Jacobi considered himself Kant's ally in all attempts to disprove the notion that God's being could be proved rationally. Mendelssohn, although he had not yet read the *Critique of Pure Reason*, knew that Kant did not share his views about rational proofs for the existence of God; he also knew that the irrationalist Jacobi had even less reason to call on the Königsberg philosopher for support. Therefore he appealed to Kant's scientific impartiality and begged him to take a position in the dispute. There were other similar appeals, and so Kant was literally dragged into the fray.

Mendelssohn's stand was naturally closer to his, Kant's, own: he himself was a product of Wolffianism. Even when he had realised the limitations of enlightenment he had remained true to its best traditions; he criticised the doctrine but did not repudiate it. Not long before the beginning of the "pantheism conflict" Kant had published a panegyric to the intellectual movement out of which he had grown: *What is Enlightenment?* The answer to that question was simple, Kant thought: enlightenment was man's departure from an immaturity for which he was himself responsible, *viz.*, from the inability to use his reason without someone else's guidance. "Have the courage to use your own reason!" was the motto of Enlightenment.

However, it had also been clear to Kant for a long time that man was not only understanding, or a sum of scientific knowledge. As for Mendelssohn, Kant was firmly convinced that his attempts to prove the existence of God and the immortality of soul by logical means had very little to do with science. To demonstrate his disagreement with Mendelssohn it was enough to quote a few paragraphs from the *Critique of Pure Reason.* Kant did just that in a brief foreword to the book by one of his followers, *Magister* L. Jakob, of Halle: *An Examination of Mendelssohn's Morning Hours.* To clarify his relationship with Jacobi, Kant had to write another work. It was entitled: *What Does It Mean: to Orient Oneself in Thinking?* and appeared in *The Berlin Monthly* in October 1786, fast on the heels of Jacobi's book.

Compared to the *Critique of Pure Reason* this essay contains some new thoughts. If in the critique Kant sought to discredit attempts to prove the knowledge of God by means of logically consistent statements, in the treatise he denied the possibility of an intuitive, unmediated knowledge of God. Kant writes: even if in all whereby God offers himself directly to man there were nothing that contradicted the knowledge of his being, such appearances would not be proof of his being. "No one can become convinced of the being of the highest essence by appearances." Man must not confuse knowledge and faith. Kant, to be sure, is not against faith. Faith is a compass, a lodestar. To find one's bearings in thinking is to be lead by a subjective principle in the pursuit of truth if there are no objective principles available. Immediately the objection arises: the knowledge of objective truth is then superior to faith. Kant says "yes," but at the same time, "no." There is, as he qualifies the statement, rational faith, which is identical with morality. This rational faith is a "conviction (if everything in man is morally good) which is not inferior to knowledge even though it is essentially different from it."

The essay *What Does It Mean: to Orient Oneself in Thinking?* was respectful in tone and did not distrub Kant's relationship with Jacobi in the least. They continued to exchange letters, works, and compliments.

The conflict with Herder had taught Kant to be restrained in polemics and to spare the opponent's vanity.

Kant's critique of Jacobi brought the first stage of the debate about pantheism to a close. Then there was an unexpected development: Herder entered the polemic with *God. Conversations* (1787); his work on *Ideas* had for the time being been set aside. Herder attacked Kant indirectly by taking up the defence of Spinoza's doctrine so as to show that it was totally compatible with Christianity. That provoked protests from the radical Spinozists who — quite correctly — saw that the Weimar superintendent had taken liberties with Spinoza's texts. The debates continued for several years: they now dealt directly with Spinoza and pantheism.

But Kant no longer allowed this to upset him. He had decided that he was not going to be drawn into polemics. He announced this solemn promise in the foreword to the second edition of the *Critique of Pure Reason* (1787). At this time Kant could no longer complain that his *Critique* was unnoticed. But it was misunderstood. Statement after statement, nay, book after book came out against Kant. They accused him of scepticism and subjective idealism. The physicist and philosopher Lichtenberg, himself not a Kantian but a very sharp and just man, thus sketched the logic of attacks on Kant: "If Kant is right, we are wrong; but since that simply cannot be, since there are so many of us so intelligent, so learned and so upright, it is as clear as day that Kant is wrong."[126]

It was not possible to answer all, nor was it necessary: as the number of critics grew so grew the number of followers who were ready to defend Kant's interests with the pen and with the sword. That is not an exaggeration: in the spring of 1786 Kant's philosophy was the cause of a duel in Jena. A student said that in order to understand the *Critique* correctly one should spend at least thirty years in the university; another student challenged him to a duel forthwith. The "Kantian fever" spread throughout the universities of Germany. In some places the authorities became uneasy. In Marburg the count forbade the teaching of Kant's philosophy with the explanation that it subverted the foundations of human knowledge.

Kant was elected *Rektor* of the university (this office was held for one year), and the Berlin Academy of Sciences made him a full member. But Kant was not only recognised by the academic world. In *The Teutscher Merkur* there appeared a series of anonymous *Letters on Kant's*

Philosophy. They were lucid and written in a popular style; they drew enormous attention. Their author was Kant's erstwhile opponent, K.L. Reinhold. When he had finally read the *Critique of Pure Reason* he was in a tremedous upheaval. In a letter to Kant he admitted to having written the anti-review of Kant's review of *Ideas for a Philosophy of the History of Mankind*. Now he assured Kant of his affection, thanked him for the beneficent revolution that Kant's ideas had wrought in his soul, and promised to dedicate himself to their defence and popularisation.

Kant himself rarely answered his opponents. A scientific debate was interesting to him only if it brought forth new views. To show the opponent up as a fool was a thankless task, especially if there was new work waiting to be done. A broad unexplored continent of knowledge opened up before Kant, one which great philosophers had long sought to circumnavigate and where he himself had been trying to go for a long time.

The issue was ethics. Here Kant's contributions are no less great than in epistemology. We already know that it was Kant's interest in ethical problems and the difficulties he encountered in solving them (especially the antinomy of freedom) which prompted him to begin work on the *Critique of Pure Reason*. Upon the completion of his major work he returned again to ethics as such. The first harbinger of the new work was a review of Johann Schulz's *An Introduction to the Theory of Morality*. (The review appeared in a Königsberg paper before Kant's decision to avoid polemics.)

Schultz and Schulz are very common German names, and several bearers of the name played notable roles during this period. The reviewer of Kant's book was a colleague of Kant's in Königsberg; he is not to be mistaken for the already mentioned friend of Kant. There was also "Zopf(= Queue)-Schultz," a prominent and fruitful writer and critic, a freethinker at the altar. He refused, for example, to face his congregation (at Gielsdorf, near Berlin) wearing the traditional wig; rather, he wore his hair in a fashionable queue and so they called him "Queue-Schultz". He endeavoured to convince monarchs of the benefits of atheism, claiming that religion was a weak support of the throne and that an atheist made a better subject than a believer. The Brandenburg consistory called the freethinker to account, but Friedrich II took him under his personal protection. The king's favourite saying was: "Think whatever you like and as much as you like, but obey!"

"Queue-Schultz" took an active part in the debate about pantheism. He lectured Mendelssohn because Mendelssohn did not want to concede that Lessing had been a Spinozist. Spinoza's followers are called atheists, so what? It is a widely held opinion that an atheist is a dangerous, amoral man, fit to be neither the father of a family nor a citizen of the state; but there is nothing more nonsensical than such statements. Morality and religion are two different things.

In 1783 "Queue-Schultz" published a book on morality, which Kant reviewed. The book laid out the foundations of mechanistic materialism, and Kant gave a conscientious exigesis of the issues. Schultz believed in a universal natural soul. There was nothing inanimate in nature; there was only life endowed with a greater or lesser degree of reality. Soul as something different from body is a product of the imagination. All beings are mechanisms. Both the seraphim and the tree are artful machines. Consequently, there is no free will: everything is subject to the rigid laws of necessity.

The last statement intrigued Kant greatly and to it he directed his critique. Fatalism, in his considered view, converted every instance of human behaviour into a puppet-show and completely eliminated the duty of moral behaviour. The latter however is for Kant the corner stone of ethics — the practical application of reason.

The first systematic exposition of ethics was undertaken by Kant in the *Foundations of the Metaphysics of Morals*, published in 1785. Why did Kant not call his work *Critique*, analogously to the *Critique of Pure Reason?* He explained it by saying that the issues in ethics were not as complicated as those in epistemology; here reason did not have as many dialectical traps to face as in the realm of theory; understanding of the most ordinary kind could easily reach a very high level of veracity even without an application of critical apparatus. On the other hand, Kant was convinced that such a critique could only be brought to completion when it became possible to demonstrate the unity of practical and theoretical reason (i.e. morality and science). In 1785 Kant did not yet think that he was ready to solve a task of such magnitude. As soon as he felt up to it he began work on the *Critique of Practical Reason*. The book came out in 1788. The content of these two books on ethics is cross-referential, as it were: in part they reiterate each other's points, in part they amplify them.

These books contain only the beginnings of Kantian ethics; its complete form will only be realized in the later works. Kant had spent many long years on the theory of cognition; as a result it was produced as a whole, was laid out in a severe, organic, systematic fashion. With the theory of morals it seemed that matters would be much easier, but it turned out the other way: in his ripe old age Kant finally produced a work in which every thing had been thought out to the end — *The Metaphysics of Morals.*

To human behaviour Kant applied a new term: the autonomy of morality. Previous theories had been heteronomous, that is to say they had derived moral propositions from principles extraneous to morality. Some saw the source of moral principles in some compulsive law — the will of God, social regulations or the demands of innate feelings. Others claimed that our ideas of goodness and evil were derived from desired human goals, or from the results of human behaviour, or that they were derived from the striving for happiness, pleasure, and utility. In contrast to this, Kant established the fundamental independence and intrinsic merit of moral principles.

The fundamental concept of Kantian ethics is the autonomous good will.

When Kant speaks of it, he does so with great pathos:

> It is impossible to conceive anything at all in the world, or even out of it, which can be taken as good without qualification, except a *good will.* Intelligence, wit, judgment and any other talents of the mind we may care to name, or courage, resolution, and constancy of purpose, as qualities of *temperament*, are without doubt good and desirable in many respects. They can also be extremely bad and harmful when the will, which has to make use of these gifts of nature and which therefore has the term "character" applied to its peculiar quality, is not good.
>
> . . . Even if, by some special disfavour of destiny or by the niggardly endowment of step-motherly nature, this will is entirely lacking in power to carry out its intentions; if by its utmost effort it still accomplishes nothing, and only the good will is left (not, admittedly, as a mere wish, but as the straining of every means so far as they are in our control): even then it would still shine like a jewel for its own sake as something that has its full value in itself.[127]

They say that the road to hell is paved with good intentions. But the Kantian good will is not passive: from the bearer of this good will activity and the ability to act are demanded, as well as the capacity to put every means in one's power to a proper use. Kant has been criticised for the formalist approach of his ethics: what in certain circumstances can turn out to be beneficial can be harmful in another different circumstance. That was true, and the philosopher knew it. In the introduction he only speaks of the compass which helped man get his bearings in the storms of everyday life. Naturally, each compass is subject to error, but errors pass and the needle returns to point true again: in the same way the loss of moral orientation is temporary; sooner or later the moral horizon opens before man and he sees the consequences of his behaviour — for good or ill. Good is good even if no one is good, the criteria here are absolute and evident, like the difference between the left and the right hand.

The ability to tell good from evil does not require a special education: intuition is enough. We already know that Kant preferred not to use this term; his term is "practical judgment"; it comes from "God", from nature and not from knowledge; "we need no science and philosophy to know what man must do in order to be honest and good, as well as wise and benevolent."[128] Here Kant parts ways with the father of ethics — Socrates: for Socrates the good is identical with knowledge and the absence of knowledge is the sole cause of all moral imperfection. Kant — although himself the child of the Enlightenment and its fierce proponent — leaves the frontiers of enlightenment rationalism behind. Science and morality are different spheres of man's being. They are connected, yes, and he will come back to this; but first he is interested in the differences.

When reason abandons the empirical and enters the realm of pure theory, it becomes self-contradictory; it encounters enigmas, the chaos of ignorance, uncertainty, and instability. It is different with practical conduct. When the practical capacity for judgment frees itself of sensual material, it sets aside extraneous circumstances and simplifies its task. Morality thus shines forth in its pure, unalloyed form. It is in this way that philosophy cooperates with morality although morality's source is outside of philosophy. "Innocence is, after all, a wonderful thing, but it is also a great pity that it cannot be preserved easily or recovered without pain. Therefore wisdom itself — which consists more in activity than in knowledge — needs science, not in order to learn from it but in order to embody its injunctions and give them power and duration."[129]

Reason is constitutive only in the sphere of the practical (the moral): there it fulfills the task of constructing fundamental concepts and realizing them. (We recall that in the sphere of knowledge reason is regulative, i.e., it only eliminates error; only the understanding is constitutive in knowledge.) The object of practical reason is the greatest happiness, i.e., the revelation and actualization of what is necessary for man's freedom. Kant speaks about the primacy of practical reason over theoretical reason. Practical conduct is what is important: in the beginning is the deed, cognition follows. Philosophy here frees itself from the bonds of speculative constructions and enters the sphere of the important problems of practical life in so far as it thus helps man find a firm foundation for morality.

Philosophical analysis of ethical categories does not precede from experience: categories are given *a priori* in human reason. Kant reiterates this notion in several places. It must be understood correctly. He does not investigate the source of ethical categories as forms of consciousness which come into being together with social relationships and undergo change together with them. The only issue here is the moral status of the individual. The daily experience of antagonistic society stands in opposition to morality; this experience perverts the mind rather than educate it. Moral conduct is the result of an inner imperative (a commandment) which at times runs counter to the amoral praxis of the environment.

Strictly speaking, each action is imperative: in order to come to pass it requires the concentration of the will. In Kant's view, however, it is necessary to make a distinction between imperatives bent on the accomplishment of a specific purpose and those which are not determined by such a purpose. The first he calls hypothetical (the action is determined by its purpose and is therefore a means to something else); the second he calls categorical. The moral deed is the result of a categorical imperative; man does not wish to achieve any purpose by its application; the action is necessary in itself.

The hypothetical imperative can have two goals. In the first case man has a clear notion of what he wants, and the issue is simply how he goes about actualising his intent. If you want to be a physician, study medicine. The imperative takes the form of a rule of skill. The rule does not state the value or appropriateness of the set goal, it only asks how one is to go about achieving it. The prescriptions of a physician who wants to heal the patient, or of a poisoner who wants to kill him, are here of equal status; they are the means whereby the planned deed is actualised.

In the second case there is a purpose, but a vague and indistinct one. The issue here is the happiness of man. The hypothetical imperative takes the form here of advice for prudent behavior. If it were possible to articulate a clear understanding of happiness, such advice would take the form of rules of skill. Unfortunately, however, this is not possible. Although every man strives to achieve happiness, he cannot say with certainty, and in full concord with himself, what it is that he wants, what it is that he needs. Man wants to be rich — what is the price he will have to pay in consequence, in terms of worry, envy, and hatred? He wants knowledge and insight — does he need them, will they bring him satisfaction when with a sharper gaze he sees the misfortune which had remained hidden until now? He dreams of a long life, but who is to say whether it will not be one long period of suffering? He at least wishes to be healthy, but how often has a weak body protected one from dissipation, etc. Concerning happiness no imperative is possible which could exactly prescribe those actions which would make man happy; for happiness is not an ideal of reason but an ideal of imagination and rests on strictly empirical grounds.

Morality cannot be built on such sandy soil. If everybody strove to achieve only his own happiness, the maxim (the rule) of human conduct would assume a very peculiar "universality." A "harmony" would ensue, such as was set forth by a satirical poet: a heartfelt harmony of the married couple who destroy each other: what a wonderful harmony! That which he wants, thus wants she! Under such conditions it is impossible to define a moral law which would apply to all.

Matters do not change when man elevates universal happiness to the level of a principle. Men cannot reach an agreement about this either, the purpose remains vague, means are uncertain, everything depends on opinion, and that is unreliable at best. (Therefore no one can force another to be happy as he wants them to be happy, according to his own notion of happiness.) It is possible to consider moral laws as objectively necessary only when they apply to all and rule reason and will.

Kant's categorical imperative in its definitive form runs as follows: "Conduct yourself such that the maxim of your will could always at the same time serve as the principle of general legislation."[130] Actually, this is a paraphrase of an old truth: treat others as you would want them to treat you. Do that which everybody should do.

It is easy to criticise Kant's categorical imperative: it is formal and abstract, like the biblical commandments. For instance: thou shalt not steal. What if I am dying of hunger and I steal a piece of bread, and my neighbour will not suffer want thereby? Kant does not want men to go hungry while food goes to waste. He only wants to call things by their proper names. In the last extremity — steal, but do not call it moral behaviour. That is the sense of the matter. Morality is morality, and theft is theft. Distinctions must be precise.

Among Kant's works there is a small treatise with the revealing title: *On the Supposed Right to Lie Out of Kindness*. The philosopher insists that man must always speak the truth. Even if a murderer determined to kill your friend asks you whether the victim is at home, do not lie! You have no guarantee that your lie will save him. If you answer the murderer truthfully that the victim is at home, and your friend has, unbeknownst to you, left his house, the deed will not be done. If you had lied and said that your friend was out, and he had without your knowledge gone out indeed and the murderer had met him in the street and killed him, then you could be rightly considered responsible for his death. If, on the other hand, you told the truth, in so far as you knew it, it would have been possible for the murderer to seek out his victim in his home and for neighbours to have come to the victim's aid and prevented the murder. Truthfulness is an obligation; let man make the smallest exception to the rule and the whole edifice crumbles to the ground. A moral injunction knows no exceptions.

Yet exceptions trouble Kant. In his late work on ethics, the *Metaphysics of Morals,* there are many sections to which peculiar supplements (as antithesis is to thesis) are given, all with the title: *Casuistic questions.*

For example, the thesis "suicide is immoral" is brought up. The author immediately offers an antithetical question. Is it suicide if one goes to a certain death in order to save his country? Is it permissible to avoid an unjust death sentence by a free choice of death? Can one accuse the soldier who prefers death to capture of suicide? Or accuse a patient who believes that his illness is incurable? The questions remain unanswered, but they are proof that Kant did not close his eyes to the contradictoriness of life. He only claimed that morality (as well as law) could not be applied to such ambiguities. In morality man finds a steady ground which may perhaps waver in a crisis; but crisis and norm are two different things.

Duty is the most reliable ground of morality, the sole source of the categorical imperative. Only duty, and not some other motive (inclination, etc.), characterises conduct as moral: ". . . there are some souls so perfectly sympathetic that without any vanity, or ambition, or selfish motive they take an inner pleasure in spreading joy about them; they are happy in the pleasure of others as long as that pleasure is the result of their activity. But I maintain that in such cases this conduct, however dutiful and pleasurable it may be, has no true moral value . . ."[131]

This rigorous passage invited objections and ridicule. Schiller could not restrain himself from writing an epigram:

I serve my fellow gladly, more glady if I like him,
And so I worry, and I moral or not?
There is no other way, you must seek to despise him
And with repugnance do what duty gives to you as lot.

Kant gradually softened the edge of his formulations. If at first he opposed love to duty, later he found the means of pulling them together. Wisdom and gentleness come with age. One can be wise beyond one's years. Schiller's epigram must not be taken seriously; he was after all a great admirer of Kant. But perhaps the epigram did have an effect on the philosopher. In his old age he asked himself a "casuistic" question: What is the value of a good deed if it comes out of a cold heart?

Extremes do not exclude one another. In the beginning Kant opposed happiness to the moral law; in the end — after years of contemplation — he brought them together in thought. The philosopher never wanted to rob man of his claims to a happy life. He was even prepared to recognize the concern with one's own well-being as being duty to a certain extent even if only because well-being (health, wealth, and education) created better conditions for the discharge of duty while the absence of well-being brought in its train the temptation to shirk one's responsibilities. But the striving for happiness cannot be considered a moral principle in itself. As far as the happiness of others is concerned, it is described in *The Metaphysics of Morals* as an aim and duty of man. "The perfection of the self and the happiness of others",[132] — that is the final formulation of duty. Man should not forget that conceptions of happiness vary, and that it is dangerous to force one's will upon others. Common sense and humanity must always be on guard, both are extraordinarily important to Kant.

Kant belongs among the first thinkers who proclaimed the autonomous worth of the human personality, independent of race, nationality, and class. One of the variants of the categorical imperative is: "Conduct yourself so that the humanity in you and the humanity in every other being is always at the same time considered as an end, never only the means."[133] That is a quotation from the *Foundations of the Metaphysics of Morals:* in the *Critique of Practical Reason* there is a no less decisive statement: "Man is in the order of ends an end in itself (with every other rational being), that is to say, man can never be used merely as a means (not even by God himself) . . ."[134]

Much depends upon the man himself. There is one central concept to be remembered: that of dignity. Man must know what dignity is and how to protect it. Man must never become the slave of another, must never permit his rights to be abridged in silence, must not contract debts (especially if he is not certain that he will be able to repay them), must not accept charity, and must not become a sycophant or a flatterer. Thus, says Kant, will you preserve your dignity; and he who behaves like a worm should not complain later if he is stepped on.

For the "class of thinkers" in particular, Kant formulated the following maxims: 1. Do your own thinking. 2. Think yourself into another's position. 3. Always think in harmony with yourself. Man was given his intellect so that he could use it freely, so that the horizons of his mind would be broad enough and his thinking consistent to itself.

Kant repudiates every form of fanaticism, characterizing it as the destroyer of the frontiers of human reason. Even the "heroic fanaticism" of the Stoics did not attract him. The conduct of a thinking man is governed solely by the sober recognition of duty:

> Duty! Thou sublime and mighty name that dost embrace nothing charming or insinuating but requirest submission and yet seekest not to move the will by threatening aught that would arouse natural aversion or terror, but only holdest forth a law which of itself finds entrance into the mind and yet gains reluctant reverence (though not always obedience) — a law before which all inclinations are dumb even though they secretly work against it: what origin is there worthy of thee, and where is to be

> found the root of thy noble descent which proudly rejects all kinship with the inclinations and from which to be descended is the indispensable condition of the only worth which men can give themselves?
>
> It cannot be less than something which elevates man above himself as a part of the world of sense, something which connects him with an order of things which only the understanding can think and which has under it the entire world of sense, including the empirically determinable existence of man in time, and the whole system of all ends which is alone suitable to such unconditional practical laws as the moral. It is nothing else than personality, . . .[135]

To be a person is to be free; it is the ability to actualize one's self-consciousness by one's self-governed behaviour. The nature of man is that he is free. Ethically considered freedom is not merely free choice (arbitrary) (*Willkür*). Nor is it a mere logical construct by the application of which given motives can with equal right result in different modes of conduct. I will, therefore I do this; I will and now I do the opposite. The moral freedom of personality consists in the recognition and fulfillment of duty. For itself and for other men, free will and the will subject to moral laws are one and the same.

How the freedom of man is possible in Kant's view we already know from the previous chapter. Man is a being belonging to two worlds. His belonging to the sensibly perceptible (phenomenal) world makes man the plaything of external causality; he is subject here to foreign forces — the laws of nature and the institutions of society. But as part of the intelligible (noumenal) world, the world of the "things-in-themselves", he is endowed with freedom. These two worlds are not anti-worlds, they are in a state of mutual interdependence. The intelligible world is the foundation of the sensible world.

So, too, the noumenal character of man provides the foundations of his phenomenal character. Things turn bad when the latter overwhelms the former. The task of education is to enable man to be guided entirely by his noumenal character. When he faces a vital decision he should not listen to considerations of the external order (career, advantage, and the like) but solely to the call of duty. To prevent man's doing the wrong thing he is given a conscience — a marvelous capacity for self-control.

> . . . A man may dissemble as much as he will in order to paint his recollected unlawful behavior as an unintentional error, as mere oversight, which can never be entirely avoided and consequently as something to which he was carried along by the stream of natural necessity, and in this way try to make himself out as innocent. But he finds that the advocate who speaks in his behalf cannot silence the accuser in him when he is conscious that at the time when he committed the wrong he was in his senses, i.e., he was in possession of his freedom.[136]

The mechanism of conscience eliminates the duality of man. One cannot understand everything correctly and still act incorrectly; standing with one foot in the intelligible world and with the other in the phenomenal world, knowing one thing and doing another. Man cannot play games with conscience, or make deals with it. It will not be put to sleep; sooner or later it awakens and demands an accounting.

The quintessence of Kant's severe and uncompromising ethics is plain — determine yourself, immerse yourself in the consciousness of your moral duty, be obedient unto it always and everywhere, hold yourself responsible for your deeds.

Rigorists accused Kant of inconsistency. In so far as he continued the tradition of European freethinking, Kant had also broken with the religious foundations of morality: our duty to humanity informs our moral behaviour, not God's commandments. Yet everything that Kant had posited in the *Critique of Pure Reason* as absolutely incapable of proof — the immortality of soul, the freedom of the will, the existence of God — is rehabilitated in the *Critique of Practical Reason* in the form of postulates, which although they do not enlarge our knowledge nevertheless "[give] objective reality to the ideas of speculative reason in general (by means of their relation to the practical sphere) and justify it in holding to concepts the possibility of which it could not otherwise ever venture to affirm."[137] We have already heard Heine's retort to that.

Schopenhauer also had his fun with these propositions. He compared Kant to a man who follows an unknown beauty at a costume ball in order to make her acquaintance. At the end of the evening the lady takes off her mask — and it is his own wife. Kant promised to establish an ethical system without God, but all of his verbal tricks are only masks, hiding the well-known face of religious morality.

Heine and Schopenhauer missed an important detail: for Kant religion is not the source of morality, but its consequence and support. Morality sets man apart from the beast, but whence it came remained for Kant the greatest secret of creation. Just as the creation itself remained a secret: "Two things fill the mind with ever new and increasing admiration and awe, the oftener and more steadily we reflect on them — the starry heavens above me and the moral law within me."[138]

The text comes from the concluding part of the *Critique of Practical Reason.* (It is also inscribed in the marble tablet near the philosopher's grave.) Kant says further that although wonder and awe can propel us in our search, they cannot replace it. What does man need in that search? First and foremost, a scientific method. It is just as essential to the study of the world without as it is to the study of the world within. Philosophy must always remain the guardian and teacher of science. Kant never lost the firm belief in the greatness and usefulness of the scientific task imposed upon him.

The word uttered is not always followed by the deed. It is easier to teach than it is to follow teachings. The history of philosophy contains many examples of discord between the proclaimed theory and actual conduct. Schopenhauer, who preached the joys of ascetic life, was a glutton and a *bon vivant.* Nietzsche, the creator of the superman, apparently suffered from an inferiority complex. Kant the moralist and Kant the man, however, are one and the same person.

It is true that he did not always follow the rule of the categorical imperative. He was petty (especially in his old age), quirky, impatient, sometimes miserly (even when his circumstances had improved immeasurably), pedantic (although he himself defined pedantry as an evil, a "sickly formalism," and railed against it), and bore no criticism. Life demanded compromises of him, and so at times he acted with cunning and diplomacy. But in general his conduct corresponded to the ideal of the innerly free personality which he presented in his writings on ethics. Free personality was his goal and declared duty; it was the ability to determine his desires and passions, even his whole being. It was his character and his goodness.

Nature endows man with a temperament, but he builds his own character. Kant thought that it was useless to try to gradually become better. Character springs forth in an instant, as an explosion and uprising — as a moral revolution. Men first begin to feel the need of a moral renewal in their maturity — to Kant it happened when he was in his forties. Material independence came later.

In 1784 Kant acquired his own house. Now he could easily put out 5,500 guldens for the estate of the painter Becker (who once painted his portrait). The house was in the center of the city, in a quiet side street, Prinzessinstrasse, not far from the royal castle. The house had a small garden with shady trees and many flowers. The pleasures of the garden were short-lived: youngsters began to throw stones over the wall. The police were unable (or unwilling) to put a stop to it, and so the philosopher stopped using his own garden.

He had to seek police help in yet another matter: the city jail was on the other side of the garden. In the summer, when windows were open, loud choral singing would waft in. That disturbed his concentration. Besides, the philosopher was convinced that the inmates were not interested in saving their souls but in currying favour with the authorities. He asked that the singing be done with windows closed and *sotto voce*: then the warden could hear them and appreciate their god-fearing sentiments. Kant had friends in the city hall and got his wish.

The house had two storeys and eight rooms. On the ground floor was the lecture hall, where the professor taught his students, and the cook's quarters. On the first floor were the dining room, bedroom, parlour and study. In the attic lived his servant Lampe, a retired soldier. The dining room contained a table and six upholstered chairs, a glass-case holding his modest china, and a writing desk holding his silverware and money. In the study stood two simple tables, covered with books and papers; on the dusty wall (no one was allowed to clean here) hung a portrait of Rousseau. The library had no more than 500 books, including pamphlets; it was housed in the bedroom. (For the sake of comparison: Goethe had 2,300 volumes, Herder 7,700.) The bedroom had one window which Kant kept shut at all times: that, he thought, was the best way to keep out insects. Lampe had to air the room in secret, when the master of the house was not there. The bedroom was not heated even in the deep of winter. When the philosopher worked, a deathly silence reigned in the house.

Kant's day began at five o'clock. (In those days before electricity, early rising was quite usual.) Kant counselled his young friend Kiesewetter to rise at four (the Russian governor of East Prussia, Vasily Suvorov, the father of the generalissimo, was up by two). At a quarter of five Lampe appeared in the professor's bedroom and did not leave until the professor was up and about. Wearing his dressing-gown and nightcap Kant would go to the study, where he drank two cups of weak tea and smoked a pipe, the only time he smoked during the day. (Tolstoy was wrong to accuse Kant of a partiality for tobacco; he thought that if Kant had not smoked so much the *Critique of Pure Reason* would not have been written in "so unnecessarily incomprehensible a language.") Kant loved coffee but carefully abstained from drinking too much; he considered it harmful.

The first hour of work was the most fruitful and satisfying of all. If he were to lecture that day, the second hour was devoted to preparations. Classes usually started at seven o'clock; Kant now lectured nine hours a week and as a rule offered logic and physical geography in the summer and metaphysics and anthropology in the winter. After classes the professor changed back into his dressinggown and worked in his study. At a quarter to one he changed again. Friends who had been invited to dinner would begin to arrive. The old ones — Green and Motherby — were joined by the more recent ones: the military councillor Scheffner; the court chaplain Johann Schulz; pastor Borowski; Kraus, Rink and Jäsche, all philosophers; Gensichen the mathematician, the writer Hippel, Jachmann the physician, the theologian Hasse, and others. The host himself handed the silverware over to the servant. On the dot of one Lampe appeared at the door and pronounced the sacred formula: "The soup has been served." The guests filed in the diningroom and quickly sat down: they knew that the host was hungry.

Kant never ate alone. The philosopher thought that one should never eat alone, for this did not restore one's strength but rather exhausted it; when one eats alone one remains with one's thoughts and work never ceases. The renewal of one's spirits can only be accomplished by the easy conversation which flows back and forth among guests who join one for dinner. In order for everyone to take part in the conversation there should not be too many guests at the table: according to the considered opinion not more than the number of the muses and not fewer than the number of graces. Kant held here with the golden mean: he had only six place settings.

Conversation at the dinner table is a great art: one must be able to speak to all the guests, not only to one's neighbour (it is not good when the company breaks into small groups); long, embarrassing silences must be avoided (although short pauses are permissible); it is not good to jump from one topic to another (if the topic has been exhausted one should quickly introduce the next one); passions must not be fanned in society; a dinner conversation is a game, nothing more; if a serious conflict arises it must be pursued with attention and a respect for the opinions of one's companions. It is best to end a debate with a joke: it not only calms the opponents but by eliciting laughter it helps the digestion.

This was Kant's understanding of the "aesthetics of conversation," to which he attached great significance and in which he was, as a host, a great master. Jachmann noted that one came to know Kant only in the circle of his guests. As time went on his knowledge grew and grew, and he could speak about any topic with animation. To his natural wit and a love of company were added the skill of his profession to command the audience, to attend to his interlocutor, and to desire to be interesting and understanding to the latter.

Those who were used to being the center of attention could of course have been displeased by Kant's way of becoming the soul of a conversation. The young count Purgstall, who had come to Königsberg with a letter of introduction from Reinhold, was put out by the host's tendency to avoid learned themes and prefer witticisms and jokes. According to the count Kant spoke throughout the dinner, "he knew better than I what kinds of birds we have at home (i.e., in Steiermark), what the country looks like, how educated the Catholic clergy are, and so on. He contradicted me in all these things."[139]

Dinner was the only meal the philosopher permitted himself. It was fairly hearty, with good wine (Kant did not like beer), and lasted until four or five o'clock. "He ate not only with a good appetite, but sensually," reported one of Kant's guests (who probably did not know that by dinner time the philosopher had eaten nothing for twenty-four hours). "The lower part of his face, the whole periphery of his jaws, gave an impression of unmistakable pleasure; the expressive glances he shot now at one dish, now at another, bore witness to the fact that in that moment he was a man of the table."[140] Kant liked to eat well, had a keen understanding of the preparation of food and could hold forth on that subject with ease. Hippel claimed that Kant was planning to write the *Critique of the Culinary Art*. His favourite dish was fresh codfish. After dinner the philosopher went for a walk. While Green was alive (he died in 1786) Kant usually went to his place and they slumbered in easy chairs; now he considered sleep in the middle of the day harmful and made certain that he never fell asleep. Then came the time for the legendary walk. The Königsbergers were used to seeing their celebrity every day, quietly walking along the same path — the philosopher's path — usually alone, his head bent by age and thought. The wig no longer sat on his head just right, as it used to in his youth, it now slipped to one side. While he walked Kant tried not to think, but thoughts came and then he would sit down on a bench to write them down. His walk ended at the Friedrichsburg Fort. When he returned home, the philosopher attended to the household matters. Evenings were spent with light reading (papers, journals, *belles lettres*); if anything occurred to him, he would set it down on paper. At dusk, before the fire was lit, his gaze would turn to the spire of the Löbenicht church. He got so used to the sight that he became upset when the poplars in the neighbour's garden hid the church from view; he regained his peace when the neighbour acceeded to the philosopher's entreaties and regularly cut the tops of his trees. At ten o'clock in the evening Kant's house sank into a deep sleep.

The regularity of life and the strict observance of his rules of hygiene had but one aim: the maintenance of his health. Even during the years of work on the *Critique of Pure Reason* there had been cause for alarm; complaints of indisposition and quickly becoming tired appeared in letters. Before his eyes stood the depressing example of the prematurely senile Mendelssohn. Kant put no trust in medicines, he thought they were likely to damage his frail nervous system. At one point he was regularly taking preparations for excess stomach acid, but only felt better when he stopped taking them. And so, at the threshold of old age, Kant imposed an even more rigorous regimen on his life; he had developed the regimen after much self-observation and self-suggestion. It was a unique hygienic experiment, and it is interesting even today. Men who have nothing to do with Kant's philosophy still study his way of life.

For example, the Soviet writer Mikhail Zoshchenko. (He did a piece of psychological research that drew high praise from the famous psychologist I. Pavlov.)

In his story "The Return of Youth" Zoshchenko writes about Kant: "He had studied his bodily construction with great thoroughness — his machine, his organism — and he had observed it like a chemist who observes a chemical reaction as he adds now one, now another element.

And this art of prevention, of maintaining life, protecting it and prolonging it, is based on pure reason alone.

With the force of reason and will he prevented a whole series of symptoms from breaking out with which he had at times been afflicted.

His biographers assure us that he even succeeded in developing immunity against the common cold and catarrh.

One could say that his health was his own, well thought-out creation. He believed that the psychic power of will was the supreme ruler of the body.

For the author, such a life, which more resembles the working of a machine, cannot be considered an ideal. Nevertheless it must be admitted that the experiment was successful: a long life and an impressive capacity for work bear brilliant witness to that."

Kant has left an account of his "system." (Here we are speaking of his system of health, not his system of philosophy.) It is to be found in *The Strife Among the Faculties*, whose third part is entitled "On the power of the spirit to master illness by sheer intent." The work was published towards the end of Kant's life, but he had already formulated its essence in 1786: rule your nature or else it will rule you! Kant knew that his prescriptions were thoroughly individual. ("Each man has his own method of maintaining health, and any deviation from it is dangerous," he declared in one of his letters.[141]) His prescriptions were thus not patterns to be followed blindly, but invitations to think and develop one's own norms of conduct. From the standpoint of modern medical practice several of Kant's recommendations are untenable.

The fundamental rule of dietetics (as Kant calls the art of prevention, in contradistinction to therapeutics — the art of healing) is not to be indulgent with one's powers, i.e., not to weaken them by comfort and indulgence. To abstain from exercising an organ is just as harmful as exercising it excessively. The motto of the Stoics is "sustine et abstine": one should be guided by it in health as much as in virtue.

Kant's hygienic programme is quite simple: 1. Keep your head, chest, and feet cold. Wash your feet in icy water ("so that blood vessels far from the heart will not be weakened"). 2. Sleep less. "The bed is the cradle of many illnesses."[142] Sleep only at night, and let the sleep be short and deep. If sleep will not come, it must be called forth. For Kant the word "Cicero" had a magical, sleep-inducing property; when he spoke it he dispelled his thought and quickly fell asleep. 3. Exercise as much as possible: attend to your own needs and take walks regardless of the wealther.

Kant recommended that liquid foods (soups, etc.) be avoided and that the intake of alcohol be limited. How often should man eat each day? We already know Kant's staggering answer: once a day! In mature years man can (but does not have to) check his appetite at dinner and satisfy it finally at supper. In old age the rule is different: the stomach is not sated by the first meal and must be offered another one.

It is harmful to think hard during meals and during exercise. Man should not make the head and the stomach work at the same time; the same holds good for the head and the legs. In the first instance the result is hypochondria, in the second, vertigo. (Kant knew hypochondria well: ever since his childhood he had suffered from "self-induced torture," when life is a bore, when one discovers all manner of diseases, culled from medical books, in one's organism. In this the physician is powerless; only self-control can provide cures, as Kant knew from his own experience.) The art of dietetics is actually an intelligent management of the mechanical performance of the stomach and legs together with the performance of the mind.

If man eats alone and loses himself in reading or in thought, indisposition is bound to ensue, because the work of the mind draws power away from the stomach. The same is true if one walks and thinks. In these instances the concentrating mind must give way to the "free play of imagination." That is why our philosopher always ate in the company of his friends.

But Kant did prefer to walk alone: if one walked in company one had to talk, opening one's mouth and letting in the cold air which gave the philosopher rheumatic pains. Kant paid great attention to proper breathing. It might seem trivial to us that he always breathed through the nose and held his lips firmly closed. But in Kant's day that was obviously a novelty, for he discusses this theme in great detail. Correct breathing prevents colds, fosters sound sleep, and even eliminates thirst.

Kant's ideas about the advantages of bachelorhood can make us smile. Himself a confirmed bachelor, the philosopher claimed that bachelors, or early widowers, preserved their youthful appearance longer, while fathers bore the signs of the yoke"; that gave him the idea that the former were more likely to outlive the latter. (Here Kant found opponents, who quoted statistics to the contrary.)

And finally, philosophical activity (naturally, not as a profession but as a labour of love). That is a great means for the spirit's overcoming a variety of ailments. Philosophy draws us away from outward circumstances and develops powers of the mind which counteract the infirmities of the body attendant upon increasing age. Man must keep busy; a "limited mind" must in the worst of circumstances make up something to do. An old man, for instance, collects clocks which never strike the hour at the same time, etc.

Kant refused medicines, as we already know; he looked at them with suspicion. That does not mean that he disdained medical science. On the contrary, he followed its developments and successes and showed an almost professional interest in it.

He did not tremble with anxiety concerning his life; fear of death was unknown to him. He needed his health only to be able to work; his attention to it was only a precaution necessary for the successful completion of his experiment. And the experiment was successful.

"It was a staggering experiment, and it was brought to a victorious conclusion," concluded Zoshchenko. "But therein lay an error, which made a machine out of man.

Man can, of course, develop habitual rules for his body, but it must not be forgotten that the psyche — by a ceaseless repetition of these habits — strengthens them and pushes them to an extreme . . . After twenty years Kant already had manic tendencies."

Zoshchenko wanted to eliminate everything manic from man's life so that man not become a machine (even if it is a thinking machine). The meaning of life, he thought, is not to be found in the satisfaction of one's desires but in having them (Kant thought so, too). And these desires should be as varied as possible. Zoshchenko is no doubt right. But he was dead wrong when he viewed Kant's life as tragic. "How tragic were the lives of Nietzsche, Gogol, Kant, who never knew women," we read in his *The Return of Youth*. As years went by Kant was more and more convinced that bachelorhood was right for him. The tragic can only consist in the accepted, premature loss of something vitally significant for one's own life. Kant lost only his illnesses; he lived a long life in the way that he thought necessary for him.

Certainly no mortal man has yet spoken greater words than these of Kant, which contain the core of his philosophy: Determine yourself! This great idea of self-determination is reflected back to us from certain natural phenomena, and this we call beauty.
— SCHILLER

CHAPTER FIVE

The True, the Good and the Beautiful

Towards the end of the eighties there was a new upheaval in Kant's philosophical views. On the whole he maintained the positions developed in the critical philosophy, but he refined his perspective on a series of essential issues (sometimes with decisive changes). The first of those was the problem of metaphysics. The *Critique of Pure Reason* had left the question open. On the one hand Kant had demonstrated convincingly that metaphysics was impossible as a theoretical discipline. On the other hand, he had announced a programme for the development of a new metaphysics as the science of supersensible entities (God, the immortality of soul). Kant's love of metaphysics was of too long a standing and too steady a purpose to allow a quick and painless divorce.

In 1788 the Berlin Academy of Sciences announced a new theme for a competition: whether metaphysics had made any true advances since Leibniz and Wolff. The prize went to the old Wolffian, Schwab; he asserted that metaphysics had made no advances, that, furthermore, no advances were necessary.

Kant did not take part in the competition, although he started working on a submission three times. An extant manuscript clearly shows his relation to metaphysics at that time. Kant is definitely of the opinion that there can be no theoretical knowledge of things which lie beyond the limits of sensual experience. In order to grant a concept a degree of objectivity, it is essential that it be grounded in perception (*Anschauung*). Hence we can have no theoretical knowledge of God, of freedom, or of our soul (apart from the body). In a practical sense we have ourselves constructed these objects, we believe in them and conduct ourselves accordingly. A metaphysics of the supersensible is possible only from the "practical — dogmatic" point of view. And the metaphysics of nature, in Kant's view, is only a working-up of a conceptual apparatus for the natural sciences. Metaphysics is a critique, a corrective for common sense and nothing more — so we read in the notes.

Two years before the Berlin competition Kant published the *Metaphysical Foundations of Natural Science*. If in the *Critique of Pure Reason* Kant sketched his future philosophy of nature so as to divide it into rational physics and rational psychology, he now does not consider the nature of soul a proper object of scientific knowledge. "I wish to remark that in any study of nature there is only as much genuine knowledge as can be substantiated by mathematics."[143] The soul is not an extensive quantity, and the description of the phenomenology of soul is not a natural science. The latter deals only with material bodies.

A change in a body is called movement, and movement is the proper object of a metaphysical study of nature. Kant asked the question of what we can know about movement *a priori*, that is to say before all experience, with only the help of the pure categories. Corresponding to the four kinds of categories, Kant considered movement from the standpoint of quantity, quality, relation, and modality. Thus the idea of a four-partite kinetics came into being; the first part, phoronomy, only studied the quantitative distribution of matter in space; the second, dynamics, studied matter under the influence of qualitative, effective forces; the third part, mechanics, treated the reciprocal relations between bodies and forces, and "phenomenology" was applied to the problem of motion (i.e., it inquired into the possibility, actuality, and necessity of motion). Kant's natural philosophy is of interest to the modern reader: it treats of the relationship between movement and rest, the denial of the physical reality of empty space.

Although he had now attained the position of the critical philosophy, Kant did not forget the "first love" of his youth — natural science. He continued to teach physical geography. He still showed an interest in astronomy and celestial mechanics and wrote two treatises on this theme: *On the Volcanoes on the Moon* and *On the Influence of the Moon on Weather.* In 1791 he asked his pupil Gensichen to make an excerpt from the *Universal History and Theory of the Heavens* which together with his comments was published as an addendum to the Englishman Hershel's *On the Structure of Cosmos.* (By the end of the nineties five more editions of Kant's cosmogonic hypotheses had come out.)

Kant also took an active part in the practical realization of scientific discoveries. The installation of the first lightning rod in Königsberg is connected with Kant's name. The story has a lengthy history. After the Haberberger church had been destroyed by a bolt of lightning, the city fathers asked the university for advice on how to avoid the repetition of such a misfortune. That was in 1774. The request was passed on to the professor of physics Reusch, with the stipulation that he bring in a competent specialist, a philosopher. Reusch chose Kant and asked for the latter's comments on his own observations. Kant approved of them: "The little I should like to contribute to your observations, if I may, consists in the following. The conduit must be so constructed that it conducts the matter of the elements from the metal covering of the spire, but does not draw it from the storm clouds. Therefore, it must be made without sharp projections and be attached to the rod and the copper sheeting only at the top."[144]

The problem of the lightning rod was solved on paper, but nothing further was done. Finally, another bolt of lightning struck the spire. The city fathers turned again to the men of learning. Reusch wrote another submission which Kant found wonderful. "This is the best treatment of this kind I have ever read as regards the thoroughness of the exposition, its brevity, orderliness and clarity. . . ."[145]

But the authorities could not make up their minds about the new instrument. A consultant was asked to come in from Hamburg. He made a few additions to the Reusch project ("so that their recourse to him would not appear totally superfluous," commented Kant). The authorities addressed a new request to the university: they wanted to know their opinion of the suggestions made by the consultant from Hamburg. The problem of grounding was debated for a long time: should the conduit be buried in the ground, or in the nearby body of water. Preparations took ten years. Finally, in 1784 the church got its rod atop the roof.

The attempt to introduce new looms in Prussian factories was less successful. Johann Bötticher constructed a simpler loom which produced three times the amount of cloth produced by the old one. Tests were run, and the results were sent on to Berlin with the suggestion that the inventor be rewarded, but officials there thought that the effect desired was impossible. Kant tried to help the inventor on his own. He wrote a letter to an agent he knew in Berlin, promising enormous profits that would enhance "the industry, and with it the general well-being whenever a better way of thinking is put into practice."[146] Again the answer from Berlin was negative. In England they were introducing even better looms at the time!

Kant's principal interests still lay in the sphere of philosophy. When he realised that it was impossible to resurrect the edifice of speculative metaphysics which he had destroyed, he began to look for new ways to construct a total philosophical system. For in philosophy he valued the systematic approach above all, he himself being a great systematic thinker: "Historical knowledge can be achieved without the benefit of a system. To a degree it is possible in mathematics as well, but there is no philosophy (of pure reason) without a system. The outline of the whole must precede its parts."[147]

Now the general outlines of his thought had long been established. But there was as yet no system. Naturally, both *Critiques* were integrated in a very distinctive manner; one single concept was developed in both. But the unity of theoretical and practical reason striven for did not seem satisfactory to him. A crucial connexion was missing.

The shape of a whole system first rose before Kant when he uncovered a genuine, independent world between nature and freedom — the world of the beautiful. While Kant was working on the *Critique of Pure Reason* he did not think it was possible to consider aesthetic problems from the position of universally binding theory. The principles of the beautiful are empirical in character and consequently cannot serve as the foundation of universal laws. The term "aesthetics" there designated the doctrine of sensibility, the doctrine of the ideality of space and time. But in 1787 references to a *Critique of Taste* appear in his letters; at the end of the year Kant informed Reinhold that he had discovered a new universal principle of the activity of the mind: "the feeling of pleasure and pain" (*der Lust und der Unlust*). Kant's philosophical system was beginning to take on a sharper outline.

It appeared to him to consist of three parts corresponding to the three faculties of the human mind: the cognitive faculty, the faculty of judgment ("the feeling of pleasure"), and the faculty of will. The first and the third parts — the theoretical and the practical — are dealt with in the *Critique of Pure Reason* and the *Critique of Practical Reason.* The second part Kant had until now called teleology, the doctrine of purposiveness. Now teleology had yielded its place to aesthetics — to the theory of the beautiful. Kant wanted to finish the envisaged work in the spring of 1788. But his work again took a long time. It would take two more springs before the manuscript was ready for the printer. The treatise was entitled the *Critique of Judgment.*

We are already familiar with the term "judgment" (*Urteilskraft*). In the *Critique of Pure Reason* the term designated one of the intuitive cognitive faculties. If the understanding supplies us with rules, then the faculty of judgment gives us the skill to use these rules in concrete instances; it is therefore nothing but the independent use of the power of thought. We say that it is not sufficient for the judge to know the laws; if he applies them formally he passes "legal but bad" judgments; he must judge with a sympathetic, balanced sense of what is right, taking into consideration all the circumstances of the case at hand. In folklore there is a type of simple man who always acts schematically and in consequence finds himself in curious situations. Kant would say that the latter lacked *determinative* judgment, as he called the skill to apply general laws to particular instances.

Now Kant is considering another form of intuition, which he called *reflective* judgment. One is to seek a nonformal universal for the given particular; it is no longer a question here of abstracting universal characteristics: that is the work of the understanding. If there is a green leaf before us, the understanding comes up with the concept of leaf with the help of the productive imagination having compared it with other leaves and considered it through the prism of the category of universality. If we apply reflective judgment, however, we see in the given leaf part of a development; we think about its function in the organism, we put the question to ourselves of its significance within the whole. The doctrine of purposes is teleology; therefore Kant calls this activity of reflective judgment the *teleological* one. Next to it he posits *aesthetic* judgment whereby he assumes that the artistic experience affords us the same kind of pleasure we obtain froı.. the discovery of purposiveness.

The traditional Wolffian teleology concerned itself with the wisdom of the creator who made the world with a purpose in mind. Kant takes a different approach. Purpose can be actual even though a higher reason has not set it; a harmonious reciprocity which gladdens the eye does not bespeak a creator who constituted it with forethought. Teleology is not a finalism for Kant; it is no confirmation of a divine ordering of the world. That notion was rejected once and for all in the *Critique of Pure Reason* as being an appearance which diverted from scientific knowledge: instead of uncovering the causes of appearances, the finalist merely refers to the investigation of the incomprehensible decisions of a higher wisdom.

Sharp dialectic has always whetted its knife on formalism; it is difficult to tell who enjoys priority here. "God created man so that he could be our food," as Voltaire puts the words into the mouth of the clever flea. Kant suggested another solution: parasites are an incentive to cleanliness, which in turn has the effect of preserving our health. Teleology for Kant is a principle for the observation of objects: first and foremost of the living organism, where everything is teleologically ordered. Every part here is connected with all other parts by reason of necessity, as if an intellect had put it all together with a definite purpose. In 1788 Kant discovered a sphere in the realm of human activity where the result also represents something organic: art.

In *The Teutscher Merkur* a treatise by Kant appeared under the title "On the Application of Teleological Principles in Philosophy." The treatise was part of a polemical exchange. Kant had barely made the promise to abstain from taking part in debates when he broke it and answered a critical attack by George Forster, a young natural scientist who had already gained fame. Forster became known through the account of the voyages of Captain Cook, in which he had been able to take part. In Kant's work on the races of man Forster saw an encroachment of the philosopher in the sphere of natural sciences. Kant claimed: we only find in experience that which we seek when we know what we ought to seek. Forster believed in pure experience, unfettered by any limitations whatsoever; he was against looking at things through philosophical spectacles. In his answer to Forster Kant refined his thought: "in purely empirical wanderings, without the guidance of a principle, nothing purposive shall ever be discovered."[148] As regarded the races Forster only acknowledged the existence of two — the white and the black — in addition to which he believed there to be a genetic distinction between them. That is to say, Africa had produced its people, and Eurasia had done likewise. Kant, on the other hand, maintained that there were at least four races, all of them from one single genetic origin. In the statement against Forster, Kant compared nature and art. In both instances it was necessary to see the living, organic whole. The one and only approach to living nature and the work of art — on grounds of the principle of purposiveness — was one of the main ideas of the *Critique of Judgment.* This was finally something new in aesthetics. Nature and art had been compared even before Kant, but with what results? The Frenchman Robinet, who was attracted to the idea of the living organism as a unique system, remarked ironically: works of art do not grow; they are put together part by part. Each part is complete in itself when another part is added to it. Works of art do not produce similar works of art; or has anyone ever observed that a house produced another house?

All of that was of course true, but Kant goes beyond it: "A product of fine art must be recognized to be art and not nature. Nevertheless the finality in its form must apear just as free from the constraint of arbitrary rules, as if it were a product of mere nature."[149]

This revelation shook Kant's contemporaries most profoundly. Goethe, who did not think much of the *Critique of Pure Reason*, was enthralled by the *Critique of Judgment.* This revelation has played a double role in the history of culture. First of all, the problem of artistic creation was set forth. Robinet was no doubt right when he remarked that the organism comes into being immediately as a whole, and that a work of art comes into being piece by piece — one part is already finished, while other parts exist only in the mind of the artist. But the finished product lives like an organism. And arbitrary intervention here is impermissible. The phenomenon of beauty disappears when a clumsy hand destroys the harmony, the "purposiveness" created by the artist.

In Kant's day one could hardly correctly appreciate the other aspect — the aesthetic view of nature, to which the philosopher attached great significance: "I do maintain that to take an immediate interest in the beauty of nature (and not merely to have taste in estimating it) is always a mark of a good soul . . ."[150] This interest in the beauties of nature is intellectual; we find joy not only in the forms of natural objects, but in the fact of their existence as well. Kant predicted that the future times would draw away from nature more and more. At the time it was not at all clear what he meant by that. Only in our day have Kant's views of the nature around us as a harmonious, artistic whole won a deep measure of universal significance. Nature herself is a work of art. Just as man must not interfere with the life of an artistic organism, so much the harmony of nature — the embodiment of discrete, purposive balance — remain undisturbed.

Kant's teleology is not theology, nor is it natural science: the philosopher does not use it to find God in nature, or to discover the laws that govern it. As always, man is at the centre of his observations. Only man can consciously set goals for himself, as the product of which ensues the world of culture. Kant's teleology transforms itself into a theory of culture.

There is, among Kant's manuscripts, an interesting fragment (from the mid-eighties): *The Character of the Human Species*. The philosopher asks questions and provides answers: "What is the natural end of man? The highest culture. — In which condition is this possible? Civil society. What are the springs of its progression? Unsociability and competition. Work."[151]

Kant examines two types of culture — the "culture of skill" and the "culture of discipline." The first is absolutely necessary for the achievement of goals, but it is not sufficient for choosing them; only the second frees the will of the "despotism of appetites," the bondage to things. In the broad sense of the word culture embraces everything that is contrary to nature: that which is made by man. Yet it is indifferent to the fate of man. It generates antagonisms, whose force pushes it forward. Its mechanism is inequality; one group holds another in bondage, toil, and the absence of pleasure. Kant described the condition of "alienation" (*Entfremdung*) here; although he did not use this term, he showed the contradictory influences of private selfishness and the division of labour on the fate of society. Here, progress and degradation co-exist. This outward, "technical" type of culture Kant calls civilization. To this Kant compares a more restricted sphere, which he calls the "culture of training (discipline)": it is the sphere of absolute morality. Both spheres grow and develop, but culture appears to lag behind civilization. Kant does not lose hope that one day culture will catch up with civilization. Otherwise, our prospects are gloomy.

Culture is also structurally organic; man is not only a means in culture, but also an end. Here the principle of "subject purposiveness" appears, whose indicators are the "feelings of pleasure and pain." Today the term does not seem to have been a fortunate choice: the modern reader recalls Freud's pleasure principle and thinks that Kant has an animal pursuit of satisfaction in mind.[152] Kant meant something else. There is pleasure and there is pleasure. Kant draws a distinction between pathological and moral pleasure. The latter is identical with the concept of culture. In other words: the feeling of pleasure and pain denotes, in modern terminology, a value relation. Upon it rests aesthetic judgment (artistic intuition) which makes art the point of mediation between nature and culture.

"Subjective purposiveness" is a principle of aesthetics, not of teleology; the latter (even in Kant's view) relies on objective purposiveness, i.e., on the perfection of the object. Objective purposiveness is only indirectly connected with pleasure. Naturally, it is pleasurable to see the actualization of some rational intention. But one thing is alien, an "objective" intention (of an unknown power which made nature) and the other is ours, the "subjective" human intention which is the foundation of creative culture. While working on the *Critique of Judgment* Kant narrowed down the sphere of teleology more and more, and took away its independent role; its mediating function in the system was assigned to aesthetics. Teleology in Kant's system fixes the specifications of the object and the parameters of its cognition: objective purposiveness is implied but its essence cannot be grasped. It is not sensible to establish illusory hypotheses here. In this respect teleology is analogous to theoretical reason, which slips into contradictions the moment it seeks to penetrate the essence of the thing in itself. Both teleology and theoretical reason have regulative functions. Reason plays the constitutive (i.e., constructive) role in human conduct, in morality. In knowledge it is the understanding which has the constitutive function. In the sphere of "judgment" it is the aesthetic evaluation that is constitutive, related to teleology and at the same time opposed to it.

Kant posed these aesthetic problems not because he was considering the essence of art but because he wanted to complete his philosophical system. The result is the following schema:

Faculties of the mind	Cognitive faculties	*A priori* principles	Application
Cognitive	Understanding	Laws, rules	Nature
Feeling of pleasure and pain	Judgment	Purposiveness	Art
Desire	Reason	Final purpose	Freedom

This is the final schema of Kant's philosophical system. Judgment mediates between reason and the understanding. Kant himself unambiguously designates the critique of judgment as a means which unites both parts of philosophy in a whole. That is also the title of the corresponding part of the introduction to the third *Critique*. Kant achieved a very particular victory over the dualism of science and morality, by appealing to the artistic capacities of man. The formula of Kant's philosophical system is: the true, the good, and the beautiful; grasped in their unity they are grounded in man, in so far as he is the creator of culture, which in turn is born of artistic intuition.

Kant sees aesthetics as the "propedeutic of all philosophy."[153] That means that a systematic study of philosophy must begin with the theory of the beautiful, where the true and the good first appear in its fullness. One must therefore become familiar with the third *Critique* before attempting the first two.

Let us now turn to Kant's aesthetics themselves. His predecessors, Shaftesbury and Hutcheson, emphasised the unique quality of the aesthetic, which cannot be reduced to either knowledge or morality. Kant accepts that thesis. But he also presents an antithesis: it is precisely the aesthetic which mediates between the true and the good, for in it theory and praxis flow together into one whole.

The aesthetic itself is not a monolith. It contains two hypostases, two aspects. One is turned primarily towards cognition — the beautiful; the other is turned toward the moral — this is the sublime. Kant's analysis of fundamental aesthetic categories is limited to the enquiry into the beautiful and the sublime (he treats of the comic *en passant* and does not concern himself with the tragic). That, too, is revealing: Kant is interested in aesthetics as such only in so far as it can perform its mediating function; the beautiful and the sublime are quite adequate for the solution of the task before him.

The analytic of the beautiful is constructed along the lines of the familiar four-part classification of judgment: quality, quantity, relation, and modality. The first definition sounds one-sided: the beautiful is that which pleases us without exciting interest. The judgment of pleasure stems from feeling and is bound up with interest. We judge the quality of goodness by means of concepts, and the pleasure we take in it is also bound up with interest. The judgment of the beautiful is free of sensual and rational interest. Kant had to dispose of rationalist and utilitarian constructs: that is why his formulations are so categorical. If their one-sidedness is accepted, they can be found to be the foundations of numerous formalist theories of art. They are the primary objects of Kant's critique.

The second definition of the beautiful goes further. The issue here is a quantitative characterization of aesthetic judgment. The universality of the judgment of taste is derived from this. "That is beautiful which pleases all without conceptual mediation." But if there are no concepts, how do we obtain universality? Feeling is individual, it lies at the root of pleasure and makes no claim to universality. It turns out taking pleasure in the beautiful is the result of the "free play" of the cognitive faculties — of imagination and the understanding. Herein lies beauty's "subjective universality."

Kant now asks the question of whether the feeling of pleasure precedes the agreeable judgment of the object (by means of the "free play of presentations"), or the judgment precedes the feeling of pleasure? The solution to this puzzle is for him "the key to a critique of taste."

If pleasure is the antecedent element, then the problem of universality has been reduced: pleasure cannot be passed on from one man to another. "Only cognition can be universally communicated."[154] In this case, however, there are no concepts employed in the play of our mental forces. Instead, we are in a certain "state of mind" which may perhaps be related to "cognition in general." That is the condition of a "free play of the cognitive faculties" (also "representational faculties"). The result, thanks to the free play of imagination and the understanding, is an agreeable judgment which precedes and brings forth feeling of pleasure, and gives a universal character to aesthetic judgment.

What we have before us is indeed the "key" to the problem, as well as one of the most remarkable of Kant's discoveries. He discovered that the perception of the beautiful is made possible by mediation. Before Kant it was thought (and the view can be encountered even today) that the beautiful gives itself to man directly through his senses. It was enough to be sensitive to the beautiful, to be endowed with an aesthetic sense. However, this "aesthetic sense" is a complex intellectual capacity. Even the ancients had believed that a supersensible beauty was possible. To be able to appreciate the beauty of an object one must grasp its inner value. Sometimes that happens in "a flash," and sometimes it requires time and intellectual effort. The more complicated the object, the more complex and differentiated its aesthetic evaluation. Scientific beauty is only for specialists. In order to grasp the beauty of a mathematical formula one must indeed possess a degree of aesthetic culture, but before that he must know mathematics. The universality of aesthetic judgment consists not in its direct accessibility to everyone but in its "communicability": every man can achieve it provided he puts enough time and energy in it. Moreover, man is not born with aesthetic sense, rather, he gradually develops it.

The term "free play", which Kant introduced into aesthetics more resolutely than any of his predecessors and to which he assigned a central role, must also be considered here. Every form of play sharpens the feeling of health, enhances all activity, and freshens the organization of the mind. Play is unrestrained. Play develops sociability and imagination, without which knowledge is impossible.

What is our view of play? A vital organism cannot exist in the state of passive being. If its activity is not aimed at bringing forth life, its vitality turns back upon itself. That is play as well — activity as such, without any result yet according to set rules. Play is self-contradictory: the player exists simultaneously in two spheres — the imagined and the actual. The capacity for play depends on the capacity for acting on two levels. This is the same duality present in art. No matter how lifelike an image may be, the viewer (or the reader) does not forget for a moment that he is confronting an imaginary world. If man loses sight of one of these aspects of art he finds himself outside of the sphere of its activity. Enjoying art is participation in a play. Kant grasped the central meaning of this problem.

The third definition of the beautiful brings us even closer to cognition: "Beauty is the form of purposiveness in an object, in so far as it is perceived apart from the presentation of a purpose." The qualifications attached to this definition are especially important. Besides the concept of "pure" beauty Kant introduces the concept of a dependent (*anhängende*) beauty. Flowers are the example of the former, people, buildings, etc., of the latter. Dependent beauty presupposes the concept of a purpose which determines what a thing ought to be like. That is already an antithesis.

Is the "dependent beauty" perhaps a less worthy or lower level of beauty? On the contrary. We find that the aesthetic ideal is actualised only in the sphere of the "dependent" beauty. It is impossible to imagine an ideally beautiful flower. The ideal of the beautiful, in Kant's view, consists in the "expression of the moral." One of the most important conclusions of Kant's aesthetics is: "The beautiful is the symbol of the morally good." And thus we are once more in the sphere of ethics.

Further on Kant draws us into the sphere of cognition. The issue is now at the most elementary level — that of empirical knowledge. Besides the ideal of the beautiful, Kant establishes the "idea of norm": an ideal representation of the outer form. The idea of norm is the middle ground between two given classes of appearances. If you want to know how the ideal representation of a beautiful man is determined, take one thousand representations and put them one on top of the other; the darkest area of overlapping representations is the model form. The model can also be calculated by taking the averages of individual bodily measurements. And although Kant stipulates that it is not necessary to take refuge in the actual proportions, it being possible to rely entirely on the dynamic power of imagination, he does not get any further than a completely mechanical solution to the problem, which he himself, often, and rightly, subjected to critique. (He did take this into account: in one of his later works he gave the more precise statement that average measurements are insufficient: beauty demands something characteristic.)

The fourth definition of the beautiful — "Beautiful is that which, without a concept, is recognised as the object of a necessary pleasure" — contains nothing essentially new. The judgment of taste is binding on all. Why? The condition of necessity, which is presupposed by a judgment of taste, is the idea of a "universal feeling" (*sensus communis*), which rests upon the "free play of the faculties of cognition" already familiar to us. The beautiful attracts interest only in a community; it is a means of communication, for a judgment of taste strives for everyone's approval.

The four definitions of beauty can be summed up in a single proposition: "Beauty (be it natural or artistic) is the expression of aesthetic ideas."[155] "Idea" is a well-known word: we have encountered it in Kant's epistemology. The idea of reason is also a concept to which no appearance, no representation is adequate. The aesthetic idea is a representation which "gives much food for thought," without there being any concept adequate to it: ("consequently, no language is capable of erecting and elucidating it").[155] Beauty without truth is inconceivable to Kant, but each is sharply distinguished from the other.

The mediating role of aesthetics is more crucial in the analytic of the sublime than in the analytic of the beautiful. We see this already in the notion that the beautiful is "pleasing for its own sake", and that pleasure in the sublime is impossible without "cogitation" (*Vernünfteln*). "For the genuinely sublime cannot exist in a sensible form, it is contained only in the ideas of reason . . ."[156]

When Kant compares the sublime with the beautiful, he remarks that the latter is always bound up with form, while the former can also exist in a formless object. The pleasure of the sublime is mediated. Here, there is no "play," only a "serious application of imagination"; the beautiful attracts, the sublime attracts and repels. We must therefore seek the ground of the beautiful outside of ourselves (for example, in the things of nature), the ground of the sublime in ourselves and in our ideas. What, then, is the sublime?

At first Kant gives a purely formal definition: "The sublime is that with which everything else is small in comparison", but there is immediately an antithesis: the feeling of the sublime requires an attitude of the mind which disposes us towards moral pleasure. The argument runs as follows: the perception of the sublime is always bound up with a specific inner agitation which occurs when we behold objects whose proportions or forces exceed the standards of measurement known to us. "The more terrifying the aspect the more it attracts us, if only we are safe from its onslaught; and we call these things sublime because they stretch the powers of the soul beyond what we are accustomed to and allow us to discover an altogether different kind of capacity for resistance in us which gives us the courage to pit ourselves against the seeming omnipotence of nature."[157]

The sublime is the destruction of accepted proportions, yet it has its own proportion. Kant quotes the story of the French general Savary, who in the campaign with Napoleon in Egypt had discovered that the pyramids had to be observed from a very particularly set distance. If one was too far they produced no impression; the same held true if one were too near and the eye could not take them in in totality.

The sublime is the elevating; the fearless relation to the fearful, the overcoming of fear and the moral satisfaction ensuing therefrom.

Thus the sublime is the measure of morality. It is first and foremost in the sensible representations of man, in their most extreme form — in the affects. The affect joined to the idea of the good is enthusiasm, without which "nothing great can be accomplished."[158] (Hegel would repeat the phrase.) But the absence of the affect can also be sublime "because it has the delight of pure reason on its side."[158]

In the feeling for the sublime culture plays a more important role than it does in the feeling for the beautiful. The feeling of the sublime also requires a fully developed imagination. (If the beautiful draws together imagination and the understanding, the sublime draws together imagination and reason — reason, the law-giver for conduct.)

Consequently, there is no reason to fear that the feeling of the sublime will diminish through contact with an abstract object. Imagination can make up for the lack of phenomenal aspects, in fact it can go beyond what any appearance has to offer. There is no more sublime passage in the Bible than the commandment: thou shalt not have strange idols before me. A fetish diminishes and paralyses forces. "For this reason governments have gladly let religion be fully equipped with these accessories, seeking in this way to relieve their subjects of the exertion, but to deprive them at the same time of the ability required for expanding their spiritual powers beyond the limits arbitrarily laid down for them and which facilitate their being treated as though they were merely passive."[159]

The sublime, which Kant initially considered within narrow, purely quantitative confines, opens limitless spiritual possibilities to man under the aspect of morality. The presence of the moral law in us creates the universal condition for our taking pleasure in the sublime. Kant divided the aesthetic into the beautiful and the sublime; he showed in each case the relation to the adjacent mental faculties. In the conclusion he again treats the aesthetic judgment as a whole and pulls together two contradictory definitions, this time openly characterized as an antinomy — two mutually exclusive propositions. Thesis: the judgment of taste does not rest on concepts, otherwise it would be subject to dispute (decision through proofs). Antithesis: the judgment of taste rests on concepts, because if it were not the case one could not even argue about it (claim the necessity of the opponent's complying with it).

The synthesis, as is often the case in Kant, does not come forth directly. Having established two indisputable truths he does not seek a unifying formulation but leads each of them away from the other in explaining that the term "concept" is not used here unequivocally. In the first instance, the concept is the product of the understanding; in the second, the product of reason. Kant holds that the contradiction is only apparent but that a certain result can nonetheless be achieved by it: the faculty of aesthetic judgment is directly and without mediation bound up with reason — the law-giver of morality. As pertains to the bond between the aesthetic faculty and the understanding — the law-giver of cognition — Kant denies the possibility of an unmediated relationship but allows an indirect one. The aesthetic idea "gives life" to the cognitive faculty. In its application to cognition imagination is subordinated to the understanding and circumscribed by the necessity of harmonious correspondence with its concepts; ". . . whereas aesthetically the imagination is free to furnish of its own accord, over and above that agreement with the concept, a wealth of undeveloped material for the understanding, to which the latter paid no regard in its concept, but which it can make use of, not so much objectively for cognition, as subjectively for quickening the cognitive faculties, and hence also indirectly for cognitions."[160] The formula for the synthesis has been found.

Now every thing is in its place. Each sphere of man's spiritual activity has been delineated and described specifically; simultaneously channels have been established, channels that connect each with all the others and with the central point of their confluence. The true, good, and beautiful has been grasped in its peculiar nature and yet also in its unity. All of the subtle definitions have proved to be a real dialectic.

The unity of the true, good, beautiful is also founded in Kant's theory of the arts. In Kant's aesthetics, which are rooted in a general philosophical problem, art occupies a fairly small but a very important place. All the specificities of the aesthetic shine forth here for the first time in their full brightness. It is true that the limits of art are vague, but one can retrace them with a sufficient degree of certainty. Art is not nature (it is a "bringing forth in freedom"). Art is not science. ("Only that pertains to art of which not even the most complete knowledge gives man the skill to create it.") Art is not a craft. "It is not amiss, however, to remind the reader of this: that in all free arts something of a compulsory character is still required, or, as it is called, a mechanism, without which the soul, which in art must be free, and which alone gives life to the work, would be bodyless and evanescent. . ."[161] Even here the definitions are accompanied by qualifications, without which they would ossify and lose their meaning.

In Kant's opinion, art can be mechanical (when it represents the actualization of cognition) and aesthetic. The latter, in turn, can be divided into the pleasing and the beautiful. The pleasing arts provide pleasure, amusement, and leisurely distraction (for instance, the art of "small talk", or the setting of the table). The fine arts nurture the "culture of the mind"; they provide a certain "pleasure of reflection" in so far as they draw the sphere of the aesthetic and the sphere of cognition closer together.

This, however, does not exhaust Kant's dichotomy of the concept of art. Kant gives one the first classifications of the fine arts in the history of aesthetics. The basis of the classification is the manner in which the aesthetic ideas are expressed, that is to say, how the beautiful is expressed. Different kinds of art are different kinds of beauty. They can express the beauty of thought, or the beauty of phenomenal appearance. In the latter case the artist's material is either the content or the form. Thus we have three kinds of the fine arts: verbal art, pictorial art, and the art of the play of the senses.

Rhetoric and poetry are examples of verbal art. Kant considers poetry the highest form of artistic creation.

> Poetry expands the mind by giving freedom to the imagination and by offering, from among the boundless multiplicity of possible forms accordant with a given concept, to whose bounds it is restricted, that one which couples with the presentation of the concept a wealth of thought to which no verbal expression is completely adequate, and by thus rising aesthetically to ideas. It invigorates the mind by letting it feel it faculty — free, spontaneous,

> and independent of determination by nature — of regarding and estimating nature as phenomenon in the light of aspects which nature of itself does not afford us in experience, either for sense or understanding, and of employing it accordingly in behalf of, and as a sort of schema for, the supersensible.[162]

A clever formulation, but the thought is simple: the significance of poetry lies in its perfecting our intellectual and moral potentials; in the play of thought poetry goes beyond the narrow limits of abstract conceptualization and thus strengthens the mind; it elevates man in demonstrating that he is not only a part of nature but is essentially a creator within a realm of freedom: poetry "plays with appearances of its own making without falling prey to them; it declares its own activity to be mere play, but play which can be used by the understanding for its purposes."[163] Who can now claim that Kant separates art from cognition? The only thing he does not do is claim that they are equal.

Pictorial art comprises the art of sensuous truth (the plastic arts) and the art of sensuous appearance (painting). Sculpture and architecture are plastic arts. The first produces embodiments of the concepts of things as they could exist in nature, the second is determined not by nature but by the intended purpose. The major issue in architecture is the use of objects brought forth by art; this naturally limits the effect of the aesthetic ideas. Kant also counts the applied arts as architecture. Although Kant defines sculpture as the art of sensuous truth, he hesitates to overstate the case: sensuous truth should not be taken so far as to stop being art. Kant divides painting into the art of beautiful representation of nature, and the art of beautifully arranging its products. Graphics is not an independent form of art for Kant; it is part, and a major part, of painting, because every painting must rest on a drawing.

The art of the play of the senses appeals to the eye and ear. It is the play of sounds and colours. (Kant is aware of the principal difference between them.) In its ability to excite and stimulate the soul, music takes a second place to poetry. As opposed to poetry, the impression given by music is short-lived; there is no time in it for reflection, and reason considers it less worthy than any other fine art. Kant holds that it is "deficient in civility" in that it spreads "its influence further than it is wanted (to the neighbors) and thus interferes with the freedom of others." Kant is particularly annoyed by the loud singing of religious hymns: such (largely phariseic) exercises cause much inconvenience because "neighbours must either join in the singing or stop thinking." This tirade is an echo of the struggle Kant waged for quiet in his own home. The mutual interdependence of art forms also brings forth new kinds of artistic creation. Rhetoric in combination with painting produces drama, poetry with music, song, song with music, opera, etc.

In the analysis of Kant's theory of cognition we have ascertained that he is not a formalist. He is not a formalist in aesthetics, either. That is easily seen if one will only refrain from taking propositions out of their context and does not consider theses independent of their antitheses. Here, for example, is what appears to be a categorical assertion: "In all fine art, form is the most important element." One could be satisfied simply to quote this proposition, but the proposition goes on and contains the elucidation that art may not be only a matter of pleasure, which "stultifies the mind and gradually turns the object into its opposite and the mind to unhappiness with itself." Kant's conclusion is unambiguous; the above is the fate of the fine arts when "they are not combined with moral ideas, which alone are capable of bringing forth an independent sense of completeness, fulfillment . . ."[163]

When Kant says that art is capable of describing "things which are base and disagreeable in nature" beautifully, he is not referring to any pleasure taking in ugliness or to purely formal mastery. "The furies, diseases and devastations of war, etc., can be given a beautiful representation as harmful things; they can even be represented in painting."[164] What matters is not the form, but the content, the artist's attitude, his view of life.

In order to judge works of art, one must have taste; in order to create them one needs genius. The faculties of the mind that go into the making of a genius are imagination and the understanding. Genius is nothing supernatural, mystical; it is only a unique organisation of power and originality. There are four characteristics of "genius": 1. It is the faculty of bringing forth something for which there are no specific rules. 2. Its products must be exemplary. 3. The author cannot explain to others how his creation came into being. 4. The "sphere of the genius" is not science, but art. In the realm of science, Kant maintains, the difference between a great innovator and his imitators or pupils is a matter of degree; in the arts, that difference is specific. It is quite possible to master everything Newton has set forth in his immortal works on the principles of natural philosophy; it is, on the other hand, impossible to learn how to write poetry. No Homer can show us how ideas — imagination and deep ideas — arise and come together in his head, because he himself has no idea of how it occurs.

What is the status of Kant's aesthetics today? What kinds of problems does it set before us? Where does it lead? In order to give an adequate answer we must digress a bit. Kant's aesthetics is not only Immanuel Kant, but also Friedrich Schiller.

We have already mentioned his name. It is said that Schiller was a Kantian even before he had read Kant. In fact, his dissertation at the School of Medicine of Stuttgart University was entitled: "On the Unity of Man's Animal and Spiritual Natures." The opening statement runs as follows: "Many philosophers have described the body as the prison-house of the spirit, keeping it too much in the earthly sphere and preventing it from taking flight to perfection. Other philosophers have claimed with more or less stridency that science and virtue are not so much goals as means whereby happiness may be achieved, that the perfection of man consists in the perfection of his body. It seems to me that both sides are equally one-sided . . ." Schiller believes it best to establish "a balance between these doctrinal opinions, to be all the more certain of finding the center line of the truth."[165] That is what he wrote in 1780, one year before the publication of the *Critique of Pure Reason.*

The center line of the truth! Schiller holds onto this line, and it takes him straight to Kant. It is possible to glean a critique of the Epicureans and the Stoics from Schiller's dissertation which Kant also attempts in his works on the problems of ethics. And Schiller's *The Robbers* grounds Kant's future conception of right: a just order cannot be established by violence.

Schiller became familiar with Kant's works in the middle of the eighties. The poet became a philosopher — a follower, popularizer, and expander of Kant's teachings. Schiller did not create a system of his own, but he nevertheless entered the history of philosophy. His name is mentioned together with Kant's, especially concerning the philosophy of art.

"I have indeed always wished that in order to spread these principles I might be able to find a poetic mind capable of giving representations corresponding to the concepts of the understanding; but I did not have much hope, because a talent possessing both a scholarly precision in the determination of concepts and the popularity of a vivid imagination is too rare to expect to encounter it soon."[166] Kant directed these words to the aesthetician Bouterwek, who taught critical philosophy. Friedrich Schiller deserved them more.

Letters on the Aesthetic Education of Man are based on Kant's principles, as Schiller acknowledged. Kant read them with pleasure. It is one of the most brilliant works in the history of aesthetics.

The fundamental proposition is firm: ". . . man can achieve freedom only through the beautiful."[167] (A Russian reader cannot but remember Dostoevsky's "Beauty shall save the world," a coincidence which, as we shall soon see, is not a coincidence.) What did Schiller mean?

The poet shares the Kantian conception of the contradictory development of culture. He stresses even more decidedly that mankind is drawing nearer perfection even as the restraints of the civilization press upon us more and more, and only the "balance of evil" sets limits to the pressure. The weapons of progress are antagonisms, the schism among men in pursuit of their work and their faculties. "Although the world as a whole profits by this divisive development of human powers it must not be denied that the individual caught in its vise suffers under the weight of the world purpose. Physical exercises build athletic bodies, but only a free and harmonious play of limbs constitutes beauty. In the same way, the exercise of separate mental powers creates unusual men, but only the even temperament of these powers creates happy and perfect men."[168] In other words, the division of labour, the self-imposed concentration upon a limited task, makes men masters of a craft but at the same time gives birth to what we today call the "professional cretinism" — the loss of harmony.

Art is called upon to set it all right again. The artist is a child of his age, but woe unto him if he becomes its spoiled child. So Schiller cautions the artist: "Live with your time, but be not its creation; give to your contemporaries not what they want but what they need."[169] Man needs education. The artist can and must be a teacher of beauty.

Beauty corresponds to man's hidden nature. Beauty is equivocal, as is man. As a student of medicine Schiller came to the conclusion that man is neither purely matter, nor purely the mind; as a poet and a philosopher he expressed that view more decisively. The same was true of beauty, he believed. It is both material and spiritual, objective and subjective; it is life itself and its representation. The essence of beauty is play. Therein lies its ambiguity; the real weaves itself into the fictional. Schiller had taken Kant's ideas as his own.

To be sure, the play in question is not a game of chance, where material interests rule and low passions burn. The original play is an end unto itself, it is a free activity in which man's creative nature comes forth to produce culture. Such were the Olympiads of the ancient Greece, their opposite the Roman gladiatorial contests. "Man plays only when he is man in the full meaning of the word, and he is man only when he plays."[170]

When he speaks of aesthetic education, Schiller is not only thinking of the development of the faculty for understanding art: good taste is at most only one of the components of the personality. Nor is he interested in the education of moral sensibilities for which art is to provide examples: that is not much of a task. The connexion between beauty and morality is a complicated process of mediation. Direct imitation would not lead to the intellect or to aesthetic experience.

The purpose of aesthetic education is the formation of a thoroughly developed personality, a "whole" man capable of creation as well as of consumption. It does not matter if the conditions permitting creativity are limited: the general standards of the true, good, and beautiful are not absolutely necessary. Beauty only creates the conditions for activity. Schiller who called beauty "our second creator"[171] thereby rejected a direct connexion between beauty, moral behaviour and cognition: "It has been clearly proved that beauty gives nothing to either the understanding or the will, that it takes no part in either the business of thinking or of problem-solving, that it only provides the faculty which makes both of them possible without determining how that faculty is actually used."[172]

One cannot expect that a criminal will mend his ways after reading an educational novel. Solutions to scientific problems are not to be sought in the works of art. There is no direct connexion. Yet Schiller was right when he stressed an indirect connexion. In poetic works the moral ideas are born which must then become the norms for behaviour.

An aesthetic ideal has not infrequently become the ideal of a society: the feeling of harmony desired social harmony.

In the aesthetic sphere lie both the impulse for theoretical and the impulse for practical creativity. The greatest minds of our time bear witness to the fact that intuitive flashes of insight play the most important role in scientific discoveries; intuition rests totally upon the aesthetic sensibility. Something else in this connexion is important as well: beauty is not only the repository of the creative urge, but also of the necessity to reflect upon the purpose to be attained through creating. Until recently science gave little thought to its consequences. Each addition to knowledge was seen as intrinsically good and justified. Since Hiroshima the situation has changed: the problem arose of the moral evaluation of a scientific discovery which, under the circumstances of social antagonisms, can be used to inflict harm upon mankind. It became clear that truth could not exist without the good, divorced from the criteria of morality. These criteria reveal themselves to the aesthetically cultivated man in all their fullness. Today there is a new understanding of truth, based on Kantian philosophy but not restricted by it. Truth is not only the definite knowledge of something; it is something far more than that. It is not only a correspondence between cognition and its object, but also between the object and its cognition. We speak, for instance, of a true friend and understand thereby a man whose conduct corresponds to the concept of friendship. Truth is objective, concrete; it must not be only known but also actualised. Man must create an objective, concrete world which corresponds to our concepts of it and satisfies our moral and aesthetic needs. Such an understanding of truth binds itself more closely with the understanding of the good and beautiful: the unity becomes an identity.

Kant came to aesthetics propelled by the requirements of his philosophical system, not because he was interested in the problems of art. Does that mean that his theoretical constructs in that realm were not founded upon his own experience of art? That art was alien to him and that he did not leave his mark on it? That his environment could contribute nothing to the development and satisfaction of aesthetic sensibilities?

It is true that East Prussia was not Saxony, that Königsberg was not Dresden, pulsing with artistic life; it was not the Weimar of Goethe and Schiller, nor the Jena of the stirrings of romanticism; no, not even the Berlin of the Enlightenment. As a frontier town of war-torn Baltic lands, Königsberg was envisaged as a centre of trade and a military bastion, rather than the temple of the muses. The town was architecturally insignificant. "The town as a whole has no beautiful architecture," remarked a traveller in 1800. The royal castle made an impression by its sheer size, not by any well-proportioned beauty. The same was true of the town's churches. In Kant's day gothic architecture was no longer in style. It was the time of classicism and roccoco; medieval architecture was considered barbaric. Kant's taste was the taste of his time.

He was familiar with architecture largely through illustrations and descriptions. It was by no means superficial knowledge; an Englishman who heard him describe the Westminster Bridge was convinced that Kant must have spent several years in London, and that architecture was his specialty.

With painting it was more difficult. Although Kant claimed to give it primacy over all the other representational arts, something essential escaped his attention. He called painting the art of drawing. In Königsberg he could only have seen paintings (in private collections) by Rembrandt, van Dyck, Dürer, Cranach and Ruysdael. There were no paintings by the Italian masters in Königsberg. In one of his works he confused Raphael with Correggio.

From the middle of the fifties Königsberg had had its own threatre: *Magister*Kant was one of its subscribers. Kant loved music, although he played no instrument himself. He went to concerts and preferred (especially in his later years) instrumental music to orchestral.

He valued poetry above all else, as we already know. His taste was formed in the Latin school, so it is natural that he valued Roman literature more highly than the literature of the Greeks. He thought that "Vergil had more taste than Homer." He placed the latter on a level with the judicious Wieland, in keeping with his definition of poetry as "a play of senses, ordered by the understanding."[173] His enthusiastic opinion of Shakespeare is also known. He tried to write his own verses — even if only on the occasion of the death of one of his university colleagues; several such attempts are extant. He was once even offered a chair in poetry.

Goethe was convinced that Kant had not the slightest acquaintance with his work. There are other testimonies, however, to the effect that Kant liked both Goethe and Schiller. Schiller, who thought the world of Kant, wrote to him twice: once, to ask for a contribution to the journal *Horae* which he planned and the second time when he sent him the first issue of that journal together with his essay *Letters on the Aesthetic Education of Man*. The philosopher sent a polite reply, praising the *Aesthetic Education* and calling its author a learned and gifted man, but it was not clear whether he had read anything else besides the work on aesthetics.

Kant's attitude towards Goethe could have been influenced by Goethe's sympathies for Spinozism and his friendship with Herder. In the Kant–Herder polemic Goethe remained an indifferent observer: he took no sides. In general, however, the great poet respected the "wise man of Königsberg" and considered him greater then any other living German thinker. Reading Kant, Goethe said, is like walking into a well-lit room.

Goethe knew perfectly well what he owed to Kant. In his *Recent Philosophy and Its Influence*, written after Kant's death, he gave a history of his relations with the philosopher; it begins with the first disputes over the *Critique of Pure Reason*:

> In every debate I took the side that rendered more honour to man and supported all my friends who agreed with Kant that although all our knowledge is related to experience not all of it was the result of experience. I accepted *a priori* knowledge, as well as synthetic judgments *a priori*, for I had conducted myself in my poetic and contemplative life now synthetically, now analytically, the systolae and the diastolae of the human spirit were to me like a second set of breathing orifices, never closed, ever pulsating . . . Again and again I returned to Kant's thought, some chapters I understood better than others, I thought, and so I put bits and pieces to my own use.
>
> Then the *Critique of Judgment* came into my hands; to it I owe a most joyful period in my life. Here I saw my most varied interests laid out in an orderly fashion; the works of art and the works of nature were handled in a complementary manner, aesthetic and teleological judgments threw light upon each other.[174]

In Kant's *Critique of Judgment* Goethe found ideas which agreed with everything he had yet "created, done, and thought." That was in the first instance the critique of the primitive Wolffian teleology, which had always repelled Goethe. Then there was the striving to discover unifying principles for the analysis of art and nature. Finally, there was the decided and clear exposition of the knowledge of the organic whole — Goethe's fundamental problem. Kant showed that the means for everyday understanding were insufficient here. Hegel would seek the solution to the problem in his theory of the concrete notion, by means of dialectical logic. Goethe saw another possibility. According to his theory of the primal phenomenon (*Urphänomen*), man can see the universal in the particular and discover the essence in the appearance. This view is somewhat more elevated than plain perception, but it remains determined by the sensible. Goethe called it intuitive judgment, an open reference to Kant.

In the twenty-nine volumes of Kant's collected works there is not a single quotation from *Faust*; did the philosopher ever read the work? Yet, there is in *Faust* an almost verbatim quotation from Kant: "Being is duty, however brief", as Faust tells Helen when she tries to reason about love.

Faust has been compared to the *Phenomenology of Mind*. It is true that the restless wanderings of Faust in pursuit of the meaning of life remind one of the long wandering of Hegel's world-spirit on its way to truth. But if the latter is too deeply convinced that he has found the truth, the former falls prey to his illusions. Before Faust the panorama of a gigantic edifice opened; he sang a panegyric to labour; but in the end there was nothing, only lemurs running over his grave. Did Goethe want to pass judgment on the will that acts by showing that its strivings were in vain? No, he wanted to sing its praises. But in one of his lectures Kant had said that each historical event is bound up with another, unforeseen event. It is impossible to foretell the consequences, but that does not mean that man should simply sit motionless. Therefore, "brood not"; rather live, do your duty, strive, fight, be bold — and you will be forgiven, no more. The best fortune for a historical individual is forgiveness, the worst, damnation. That is the meaning of Faust's demise. Mephisto, who believes that he holds Faust's soul in his hands, is put to shame. Love is triumphant. Goethe, who in the beginning blasphemed God, came to the apotheosis of Christian love. Kant, too, had reached this point: significantly earlier and considerably more prosaically.

In the first part of the tragedy, Faust, the individual, has what Kant would call an "empirical" character. He is prey to passions which lead him to sin. In the second part, we have a "noumenal" personality before us. He finds his way to freedom through the feeling of guilt. (The following chapter will show that Kant held the same position.) Faust is full of good intentions. He enters into the pact with evil powers not for his own gain but to benefit all. But what happens: if Faust wants to wrest new lands from the sea, the legions of the enemy will be swallowed by the waters; if he wishes to engage in maritime trade Mephisto will turn it into piracy. The results attained do not correspond to the intentions. Kant's ideas can be recognized in this work by Goethe.

Kant's writings also belong to German literature in their own right. Yet he cannot be described as a particularly good stylist. The contrary is rather the case. Kant worked very hard on his writings. He never sent anything unfinished, or half-finished, to the publishers. Even so, his major works are written in a manner which would drive the most patient reader to despair. One need only turn to any page to come across a sentence that runs for ten or twenty lines. As soon as one begins to read, one finds digressions, elucidations, stipulations in parentheses and out of them, notes in the text, and footnotes. "I would read your works more often if I had more fingers," a school friend complains to Kant. "What do you mean?" the latter asks. "My dear friend, your text is so full of parentheses and stipulations I must keep before my eyes that I sit there with my finger on one word, then use the second finger, the third, the fourth, and run out of fingers before I have come to the end of the page."[175]

This is all quite true. Yet I should like to go against the tradition and break a lance for Kant's artistic gifts, which it is true, only shine forth in a narrow and quite specific realm. He was ironic. He played with words.

Let us digress. What is irony? Let us quote A.F. Losev: "Irony consists in my saying 'yes' when I want to say 'no', with his 'yes' being simultaneously expressive of my inner, more appropriate 'no'. Let us say that only the first takes place: I say 'yes' but I think 'no' to myself. That of course is a lie, a deception. The essence of irony lies in the fact that when I say 'yes' I do not hide my 'no'; in fact I bring it more clearly to the fore. My 'no' is not an independent fact, it depends upon the 'yes' and needs it; it entrenches itself in this 'yes' and has absolutely no meaning without it."[176]

The source of irony is ancient. Philosophers had used it even before its essence was known. Socrates is seen as the father of irony, which he used as a heuristic device. With its help the Athenian sage sowed doubt concerning accepted but false apparent truths. Seemingly in all innocence, Socrates asked simple questions of his interlocutor as if he wanted to learn something from him. As a result the one who thought he knew everything became confused; irony taught man that he knew little or nothing. Irony cleared the path to true knowledge.

German romantics used irony in a different manner. For them it was not a means, but an end. If the Socratic irony was only the first step on the road to knowledge, at the end of which stood the true, good, and beautiful, for the romantics irony was sometimes the highest achievement of human spirit; once the artist-philosopher had reached its peaks he gazed at the world around him with disdain. That was the way to beauty. But what happened to the true and the good?

Even the contemporaries of the romantics recognized the weakness of such a conception: the relation to truth could then only lie between complacency and self-satisfaction. Hegel, who polemized against the romantics, demanded the return to a Socratic understanding of irony. Had he wanted to, he could have relied on Kant in doing so.

The role of the founder of critical philosophy in the development of the concept of irony has not yet been fully appreciated. Kant did not theorise about irony. He, like Socrates before him, simply used the weapon of irony to clear the ground for the foundations of his edifice of truth and virtue, and this he did with subtlety and care. Philosophical humour and philosophical irony are not available to everyone. One has to have the stature of a Goethe to grasp the ironic principle in Kant's *Critiques*. "When I mastered Kant's thought well enough to put it to use I sometimes thought that this remarkable man used irony with a devilish skill, sometimes to set the most stringent limits to the faculties of cognition, sometimes to lead the reader beyond the limits he had set so artfully."[177]

Voltaire said that the heavens had given man hope and sleep to make his fate bearable. Kant would have added laughter, but with a stipulation: if only it were possible to call forth laughter in "reasonable men", and if its necessary edge were not so rare. The philosopher cultivated his sense of humour; as years went by he took more and more refuge in it in his work, precisely in those cases in which he had to write about society, where the material itself provoked wit, where exasperating nuisances could not be overlooked, when the censor had to be deceived or when the rigid minds of the bourgeoisie had to be shaken.

Kant was quite proficient in the realm of satire. Persius, Juvenal, Erasmus, Rabelais, Swift, Voltaire, Fielding, Sterne and Lichtenberg were among his favourite authors. "Man is an animal capable of laughter,"[178] he said. He well understood the essence of laughter, even if he did not say much about it. That which causes laughter must be nonsensical, he assures us. (It is not by chance that Kant, following Shaftesbury, considered the touchstone of a theory's truth to be its capacity to withstand ridicule.) Some sort of theory is at the basis of every joke. Man expects something and then hears or sees something else, totally unexpected, and laughs. "Laughter is an affection arising from a strained expectation being suddenly reduced to nothing."[179] He illustrates his thought with three jokes. An Englishman treats an Indian to a bottle of beer; the latter is surprised by the foam which runs out of it. What is so strange about foam running out of a beer bottle, asks the Englishman. That's not what bothers me, said the Indian, what I'm surprised at is how they put it in there. An heir complains that he could not give his benefactor a fitting burial; the more he paid the mourners to assume a mournful mien, the merrier they were. Or, finally, the story of the merchant who had to throw all his wares overboard in a terrible storm, so that his wig turned grey overnight. That is all one can find about the comic in the pages of the *Critique of Judgment*. And all of this is said in one section, as a "note." Not a word about irony.

But Kant must have been a man given to irony. He lived in a divided world where attitudes, manners and injunctions forced one to say 'yes' when the voice of conscience shouted 'no'. How was it possible in these circumstances that conscience did not turn deaf, that it did not become philistine or conformist? Kant's dualistic philosophy was a peculiar search for an answer to this question: it is a philosophy which places an impenetrable barrier between the world of appearances, subject to outer necessity, and the world of "things in themselves," where freedom and moral purity reign supreme.

It must be said that Hegel saw the moral and aesthetic roots of Kantian dualism very clearly, and that he worked hard to overcome them. In his endeavours Hegel did not blunt the edge of the problems raised by Kant. The world is indeed divided; next to the original world exists an "inverted world." Tht is not merely one of Hegel's figures of speech, but a category from the *Phenomenology of Mind*. Commentators on Hegel have for some time been pointing out that this category is rooted in the tradition of satire. What is contained in the inverted world is not only opposed to the actual world. This inversion makes it possible to grasp the stealthy perversion of what goes on in the real world, as if in a false mirror. The overcoming (*Aufhebung*) of the inverted world is the overcoming of satire. Philosophy itself then reluctantly becomes its object. That was Hegel's way. In Hegel's view, the division of the world disappears into the movement of consciousness; the opposites become reconciled and blend into one another. We must remember Marx here who poked fun at one of these "movements" of the Hegelian categories; "'Yes' becomes 'no', 'no' becomes 'yes', 'yes' becomes simultaneously 'yes' and 'no'; thus opposites are placed in the balance, neutralized, and superseded."[180]

Kant had much too sober a view of things to be satisfied by any illusions. Irony stood guard and drew a clear line between 'yes' and 'no'. (The ironic man never confuses these things, his apparent 'yes' only underlines his 'no'; he can be misunderstood, but he himself always knows what is at issue.) Kant's irony was right on the mark concerning what was "inverted" in the world.

We are already familiar with *Dreams of a Spirit-Seer* — a brilliant example of Kant's ironic art. In *The Conjectural Origin of Human History* the irony is more concealed. Only a barely perceptible smile of the author draws the attention of the reader. Kant assures us that he only wishes to take "a pleasure trip" through the origins of history, using only "a holy document as a map," but his treatment of the subject is serious. The Bible becomes a historical source.

Kant presupposes an irreproachable knowledge of scripture on the part of his reader. We must first recall the text to mind. It deals with Adam's children: "And Abel was a shepherd but Cain was a tiller of the soil." Kant saw in this a "transition from the life of the hunter to the life of the gatherer of roots and fruits" — i.e., the transition to labour. ". . . 3. And in the process it came to pass, that Cain brought of the fruit of the ground an offering unto the Lord. 4. And Abel, also brought of the firstlings of his flock and of the fat thereof. And the Lord had respect unto Abel and to his offering. 5. But unto Cain and his offerings he had no respect. And Cain was very wroth, and his countenance fell." The life of the shepherd, Kant believes, is comfortable and certain. The life of the tiller of the land is difficult, subject to weather, requiring a settled abode, land, and the strength to defend it. ". . . and so the ploughman may envy the shepherd, for the heavens protect him better (V.4); but in truth, the latter is the source of trouble as long as he stays near because the grazing animal does not spare his crops."[181]

Then came the murder of the brother and the explusion of Cain to the land of Nod. It could not be otherwise, Kant thought: "for the tiller of the land was obliged to use force against conduct which others considered permissible; and if he did not want to lose the fruits of his long labours he had to leave in the end the neighbourhood of the shepherds (V.16). That exodus was the third epoch," that is to say, the epoch of the division of labour. The "pleasure trip" turned into a scientific expedition with a thoroughgoing interpretation of the first stages of world history.[182]

In the *Critique of Pure Reason*, irony penetrates deep, spreading a light there which seldom shines through to the surface. Let us remember the well-known aphorism: "I must supersede knowledge in order to make place for faith." The meaning of this sentence in the biographic context of Kant's life will become clearer in the following chapter.

How terrible is a God without morality.
— KANT

CHAPTER SIX

Faith as Hope. And Love

In the late spring of 1789 the budding Russian poet and author Nikolai Mikhailovich Karamzin embarked on a lengthy journey. In France there were the first stirrings of the revolution, but the periphery of Europe still slumbered; the young man travelled leisurely, with long stops, in the westerly direction. He kept a diary, in which he entered his travel impressions as they occurred to him. Königsberg surprised him by its sheer size. He entered the city on a market day, and the streets were swarming with noisy, festive crowds. Everywhere there were bright uniforms, light blue and dark blue, green with orange, red and white facings. Karamzin had to have his dinner with a group of officers; the conversation was about the parade which had just ended; vulgar jokes flew about, and there was uproarious laughter. The poet found the whole thing not at all to his liking: he had very little respect for the art of war and he had not come to Königsberg in order to become familiar with the Prussian army.

After dinner Karamzin set off to find Prinzessinstrasse. Lenz, the erstwhile "stormer" who now lived in Moscow, had advised Karamzin to seek out Kant. The young man entered the house of the philosopher with his heart in his boots. "I had no letter of introduction; however, he who dares, wins, and so the door to the philosopher's house opened before me. A small, slender old man, extraordinarily delicate and pale, received me. I said to him: I am a Russian nobleman travelling in order to meet great men of learning — that is why I have come to see Kant. He immediately begged me to take a seat and said: 'What I have written does not appeal to all. Very few love the deep metaphysical speculation with which I have busied myself.'"

At first the conversation dealt with a variety of topics: the philosopher displayed his brilliant knowledge of geography and history, but then the guest resolutely turned to philosophy. He was familiar with the *Critique of Pure Reason*, but had not yet read Kant's ethical works. He was beset by religious and moral doubts which he wanted to put before his host. Kant elucidated his views, speaking swiftly and softly in a way difficult to understand. The Russian had to pay very close attention. The conversation lasted about three hours.

"We are meant to act. Man can never be satisfied with what he has achieved and always strives for something else. Death stops us on the way to the fulfillment of our desires. Give man everything he wants and in the very same moment he will decide that this everything is not quite everything. Since we cannot see the aim and the end of our striving in the present, we assume a future in which these difficulties will be resolved; and this thought is all the more attractive the less there is a relationship between joy and pain, between satisfaction and suffering. I take solace in the fact that I am over sixty years old and that the end of my life cannot be far off; I hope to leave it for a better one. When I contemplate the joys I have experienced in my life I find little pleasure; but, when I contemplate the instances in which I acted in accordance with the moral law engraved in my heart, I feel the purest joy. I speak of the moral law; others call it conscience, the feeling of right and wrong. You may call it what you like, but it exists. I have lied; no man knows it, yet I feel shame. It is true that the probability of a future life is by no means a guarantee of it, but when all is said and done, reason commands us to believe in it. And what would become of us if we could see it with our own eyes? Could not its charms prevent us from the proper use of it? But if we talk about destiny, about a future life, we are thereby presupposing the existence of an eternal and creative understanding, which created everything for some purpose — for some good. What? How? — Here, the wise man must be the first to admit that he has no wisdom. Here the light of reason goes out, and we are left in darkness. Only imagination can go astray in the darkness and create phantoms."[182]

When he returned to the inn, Karamzin hastened to write down the content of the conversation. He begged Kant in thought: "Oh, worthy man! Forgive me if I have distorted your thought in these lines." The Russian traveller was impressed not only by the depth of Kant's world view, but also by another characteristic of the thinker: his tolerance, the respect he had for the views of others. For Karamzin a true philosopher is one who can live in peace with all, one who can esteem even those with whose thought he cannot agree. One must be able to point out the errors of human reasoning with noble passion, but without anger. To tell someone that he is in error and why, does not mean to crush his soul and call him a fool. Oh, under what pretences do men torture one another! Thus Karamzin summed up his conversation with Kant on the subject of religion.

Kant's philosophy of religion is intimately related to his ethics. We remember the thesis: morality is not divinely ordained. But we also know the antithesis: morality inevitably leads to religion. Human capacities are not sufficient to bring man's right to happiness and the universal duty into agreement: therefore it is imperative that an all-powerful moral being be accepted as the ruling force of the world. (Accepted? The atheist Feuerbach would have said: thought up. Kant the believer speaks far more cautiously: "It may sound problematic but it is by no means reprehensible to say that each man makes his own God . . .")[183] The tractatus *Religion within the Limits of Reason Alone* is dedicated to the grounding of the antithesis.

Kant begins with reflections on man's moral nature. There are philosophers who are convinced that man is hopelessly mired in evil. Others consider that his nature is good, and that it turns to evil only under the influence of circumstances. Both are rigorists and apodictic in their judgments. On the other side of the issue are the indifferentists, who maintain that man is neutral in nature — neither good nor bad — and the synchrtists, who consider man simultaneously good and bad. In moral matters Kant is a rigorist, but also a dialectician. Here, too, he tries to bring the opposites together — or, rather, he seeks to bring them into conflict.

Man, Kant avers, is evil by nature. There is an ineradicable disposition to him to do evil; this disposition appears to be something acquired, but it is present in man from the very beginning. On the other hand, man also has an original endowment of goodness. Moral education consists in allowing these good capacities their due, so that they may emerge victorious in the struggle with man's disposition towards evil.

A victory of this kind is possible only as a revolution in man's way of thinking. Kant had himself experienced such a revolution and was convinced that a substantive moral renewal was a rebirth of a very particular sort: it was a necessary precondition for the building of character in both the individual and the species.

There was yet another condition necessary for the victory of good: "The highest moral good cannot be actualized by the individual's isolated striving for his own moral perfection; to this end all purposes must be united in a system of good will, in which and through the unity of which it [the highest moral good] can come to be . . ."[184] The common good is a social requirement which brings forth the society as such. In Kant's religious doctrine his abstract ethics acquires a social significance.

Kant introduces the concept of the "ethical community." Without it it is impossible in the course of moralization to overcome the "state of nature," where, according to Hobbs, there is a war of all against all, where there are not only no laws, but also no moral injunctions. The church is the ethical community. A particular kind of religion in a particular stage of human spirit is an unavoidably necessary instrument in the fortification and perfection of the human community.

Even in ancient times there was an enlightened view of religion, namely, that fear gave birth to the gods. In the beginning of the Middle Ages this notion was more precisely formulated: fear established the ground, and the unscrupulous swindlers imposed this or that God on the uneducated masses. If one looks more closely at most peoples and epochs, and enquires into the religious principles that held sway in the world, it is difficult to believe that the latter are anything but the feverish imaginings of the sick. That was the conclusion of Hume, who was so highly esteemed by Kant. But why have these delusions, these feverish imaginings, taken such deep roots? How did the idea of a suffering God come into being? In Hume's opinion religion was an unfathomable mystery; the study of religion produced only doubt and the refusal to pass judgment.

German Enlightenment, which grew out of Protestantism and Pietism, always took only a historical view of religion. In his *Education of Mankind*, Lessing proposed a schema of the development of religions, based on the moral perfectability of man; looking into the future he prophesized the coming of an age without religion, when morality would reign triumphant without belief in a supreme being.

Kant went back into the past, sought the social and psychological roots of a faith in God and saw in man (in mankind) the battleground of two principles — good and evil. In the beginning evil is predominant, but the qualities of goodness make themselves felt as guilt and come to rule men.

The experience of guilt (one's own or someone else's, in which we have a share) is the foundation of morality. A clear conscience is the work of the devil, as Albert Schweitzer would finally say (he wrote his thesis on Kant's philosophy of religion). A man who believes that he is always in the right is a man who is morally lost. Moral renewal is possible only in the struggle with oneself. Looking for the roots of religion Kant suddenly discovered the sources of morality. He discovered its foundations. And he encountered the builder of these foundations whom we already know — imagination. Let us see how imagination works in the field of religion.

Fear fathered gods, and gods set prohibitions. The fear of breaking a taboo, the fear that this may already have happened, gave birth to the idea of the sacrificial offering. When the sacrificial offering turns to self-sacrifice we are in the presence of a moral and religious revolution. The man who offers himself as a sacrifice equates himself to God. Thus arose the image of the suffering "son of God," the messenger who is God and man in one.

The teacher of the Gospels declared that he was sent from heaven, proclaimed the ceremonies of the old religion of the slaves null and void and said that moral faith was the only salvation of the soul. With his death, Kant adds, his public history ends. "The ensuing history of his resurrection and ascension, which unfolds in secret and only before the eyes of his followers . . . cannot fulfill its historical mission for religion strictly within the confines of reason alone."[185]

Several years later a young, unknown *magister* of theology, Wilhelm Hegel, under the influence of Kant writes *The Life of Jesus:* a biography of a great moralist shorn of any supernatural miracles. Hegel opposes the teachings of Jesus to the teachings of Moses: the living word is set against the dead dogma, the New Testament against the Old; in all of this Hegel follows Kant's example.

The opposition of the Old Testament to Christianity is for Kant an issue of decisive significance. He sees the Ten Commandments as compulsive laws, which deal only with the outward and the formal and contain no summons to moral incentives, which is of great importance for Christianity. That is the first point Kant makes.

Moses wanted to establish only a political community, not an ethical community. "The Judaic faith in its original form is only a collection of statutory laws upon which rests the political form of the state; if any moral codicils adhere to it, either from the beginning or as the result, they are not related to Judaism as such. Judaism is not a religion, but only a coming together of a group of men who, in so far as they are all members of the same tribe, formed a political community under purely political laws, and by no means created a church."[186] This is Kant's second point.

The third point is that "Judaism excluded the rest of the human kind from its community, it considered itself a unique nation, a nation of Jehovah's chosen people who detested all other nations and were in turn detested by all. The belief in its exclusive role of the chosen inevitably resulted in their making the claim that they were the universal church of the ages."[187]

The birth of Christianity meant the complete rejection of the faith of the Old Testament. That was a religious revolution. According to Kant, the "universal history of the church" begins with Christianity.

The historicism of Kant's thinking is quite clear in his philosophy of religion. Kant sees first an original, essentially irreligious human condition; then comes the first, as yet imperfect type of religion: the "religion of service to the gods." The third stage is rational belief. The religion of service to the gods (the most widely spread type of faith) seeks to elicit favour of the highest being, which can be appeased by worship, sacramental offerings, the observation of rules and ceremonies. Man lives in the illusion that God can make him happy without his having to improve himself; it is only necessary to entreat God in a prescribed manner, by the practice of some specific external behaviour. It is in fact a business proposition, based on the principle of "I'll scratch your back, you scratch mine." The priest plays the role of the middle-man; in this kind of religion the priest performs the ritual, and the church is the temple where the ritual is performed. The religion of reason is the pure belief in the good, in one's own moral capacity, without any admixture of calculation or the transference of one's own responsibility to higher powers. It is a religion of virtuous conduct, which obliges one to seek inner perfection. The priest here functions merely as a teacher, and the church is the gathering place for instruction.

Fear was the father of the gods (and the gods set prohibitions), but then, Kant says, conscience came into play. It is the essential guiding force in the development of religious consciousness. Conscience also means "knowing with"; the image of another knowing what I think, one from whom I cannot escape, weaves itself into my self-consciousness. I have committed a transgression, there are no witnesses to it, and yet I have the feeling that there is a witness, an accuser. Conscience is fear internalised and directed against oneself. It is the most terrifying kind of fear.[188] In an institutionalised religion fear is objectified in the image of a God, who establishes laws and metes out punishment when they are broken, but whose mercy and forgiveness can be obtained. In the pure religion of reason it is impossible to do such business with God (i.e., with the conscience). Here one can only honour the commandments and follow the categorical imperative: "whatever does not serve the institution of a virtuous life, and is meant only as gesture of appeasement to God, is a religious illusion and a false service to Him."[189] That is the principle of faith of the Königsberg reformer, whom his contemporaries often compared to Luther. Although Kant did not establish a new church, he did find many followers. There is essentially no difference between the shaman from Africa and the prelate from Europe. Both strive to accomplish the same thing, *viz.*, the direction of that invisible force that shapes human destinies to their own advantage; it's only their approaches to the task which differ.

In one of the concluding sections of the *Critique of Pure Reason* Kant formulated three famous questions, which he believed to comprise all the rational interests of man:

1. What can I know?
2. What ought I to do?
3. What can I hope for?

The answer to the first question, Kant held, is provided by his theoretical philosophy, the answer to the second by the practical philosophy. The answer to the third question, which is "simultaneously theoretical and practical," is a much more difficult proposition. The *Critique of Judgment*, which became an enquiry into the problem of culture, outlined the way of hope which man should follow. Culture is the final purpose of nature, and man has been called forth to accomplish it. Can he count on external forces besides his own capacities to achieve this end? What is the nature of the hope that springs from the belief in the supreme being?

It is impermissible to hope in the assistance of supernatural forces. There are no miracles that defy the objective laws of experience, no divine secrets that defy the faculties of the mind, and no grace that ennobles our moral nature by the power of a divine authority. Kant therefore rejects prayer as a means of communion with God: if you meet a man who speaks loudly to himself you may suspect him of some slight mental aberration. He rejects church attendance as a form of idolatry, as he rejects all other ritual ceremonies. In its content religion is not different from morality; there are various kinds of faith, but there is only one religion, as there is only one morality. The belief in God is first and foremost a confidence in one's own moral strength. The second and the third questions are one and the same.

Kant does not however wish to destroy the hope in a reward after death. Conduct yourself as if there were a reward awaiting you in the afterlife. Perhaps that is not so, but never mind, conduct yourself morally! It is wise ". . . to conduct oneself as if there were an afterlife, and as if its moral constellation were the same as that which obtains in this life."[190] Religion is unthinkable without a belief in an afterlife.

God is a moral law endowed with an objective existence. But it is more. Kant is not a Stoic. The greatest good for a Stoic is an ascetic life which is departed from willingly. In Kant's view, however, suicide is the failure to do one's duty. But he does not relate hope only to the strict observance of duty (as so many familiar with his thought believe).

God is love. As the years passed by, Kant also heard critical voices accusing him of callousness; perhaps he also came to know the power of the affect that binds one man to another, the affect that binds more powerfully than fear and duty. Whatever it may have been, the older Kant grew, the more he appreciated love.

Love and duty are two different things. That was the original thesis. To say that it is man's duty to love is nonsense. When one says: "Love thy neighbour as thyself," it does not mean that you must first love man and through that love then do him good. On the contrary: be good to your neighbor and that will awake the love of man in you. To do good for other men according to our ability is a duty, independent of whether we love them or not; and this duty remains in force even when we have made the sad discovery that mankind is not worthy of love.

Thus the *Metaphysics of Morals*, Kant's last work on ethics; the antithesis is inescapable, it follows inevitably, for one of the chapters is entitled "On the Duty to Love Other Men." The reader is astounded. Then he is saved by a stipulation: in this instance love is not understood as a feeling, but as a certain universal principle.

Now we have to draw the synthesis which will take the sharp edge from the extreme positions. We find the synthesis in the treatise *The End of All Things*, where Kant considers how love can come to the assistance of the fulfillment of duty: "for what one does not want to do, he does poorly, although with sophisticated excuses from the commandment to do his duty, so that we cannot count much on the latter's incentive without the former's assistance."[191]

If man annexes another authority (even divine) to Christianity in order to make it better, be the intent itself ever so well-meaning and the purpose ever so good, its amiability will vanish: for it is a contradiction to command someone not only to do something, but to do it willingly.

The goal of Christianity is to generate and bring forth a love for the observation of duty; for its founder speaks not as giver of orders who demands subjection to his will, but as man's friend who places in the hearts of his fellows their own, fully grasped will, i.e., one according to which they would act freely, of their own accord, when they have put themselves to the test.

"It is to the free exercise of thought — as far from the mind of the slave as it is from the mind of the voluptuary — that Christianity owes the suasion of its teachings; through it it wins the hearts of men whose understanding has already been enlightened by the representation of the law of their duty. The feeling of freedom in the selection of the ultimate purpose excites in them the love of moral laws."[192] We find the same thought in the treatise on religion: "The highest aim of the moral perfection of finite beings, one which man can never fully achieve, is the love of the law of morality. Correspondingly, the principle of faith in religion ought to be: 'God is love.'"

God is love. He who has repeated this truth of the New Testament has also pondered what lay behind it. God is love, Hegel says, and sees in the statement a problem in dialectical logic: love is the identity of entitheses, its highest form is "the love of community" mediated by the "worthlessness of every particularity."[193] Love is God himself, echoes Feuerbach, but for him the word "god" is only a metaphor, he rejects Christianity in a decisive fashion. For Kant Christianity is the highest stage of the moral and religious development of mankind. Everything else is a step backwards.

But whence came love? Here, too, Kant tries to think historically. Love is a gift not from heaven but from the earth, the metamorphosis of the sexual instinct. The low animal lust, manacled by prohibitions and denied the freedom of satisfaction, transforms itself into a higher element of culture. Kant reflects upon sociogenesis.

> Man soon found that the sexual urge, which in animals is based merely on a transient, largely periodic attraction, is capable in him of being prolonged and even increased through the work of the imagination, which conducts its business with more restraint but also with greater duration and more evenness, the more the object is removed from the senses; as the result, the ennui which inevitably follows the satisfaction of sexual urges, is avoided. . . . Denial was the master stroke which converted pure attraction to an ideal yearning, purely animal lust to love, simple feeling of pleasure to a taste for beauty first only for man, then for nature herself.[194]

Kant says that this "small beginning" is more important than all of the successive achievements of culture.

The argument among Kant's followers was not only concerned with man's unconscious animal instincts. What came first — fear or prohibition? According to Kant imagination came first; it instilled fear and inhibited activity. Thus passion became purer, more powerful. Perhaps this was an inevitable conclusion given the limitations the philosopher had been forced to impose upon himself.

When he interprets the Bible, Kant names another two powerful stimuli of cultural development in addition to the instincts for nourishment and love. The expectation of the future — the thought of living for the coming generations, the hope for a better tomorrow (now no longer for one's self but for one's children) — and, finally, the need to be an end in oneself (and not a means for others).

Kant stands before us in a new and unusual hypostasis. Before we saw in him the man of irony, now we have before us an apologist and analyst of love. He understands Christianity as a moral principle, as a programme for loving one's fellow man. In ever perfecting this programme, he also seeks to give it a theoretical foundation. He is far from churchly dogmatism. He converts dogmas into hypotheses. "God as a hypothesis,"[195] as he says in his "Reflections." And further: "God is not a being outside of me, but a thought in me."[196] Doubts did not abandon Kant. At the same time he was not a fighter in God's cause.

By the mid-seventies Kant had developed a moral interpretation of religion, as his letters to Lavater show. He came to think, with Job, that it is a transgression to "flatter God."[197] To bend one's knee or prostrate oneself in order to show one's submission to the heavenly powers is contrary to human dignity, just as it is demeaning to try and enter into relationships with Gods through graven images. When man does this he does honour not to the ideal brought forth by his reason, but to an idol, a thing of his "own making."[198] Kant stopped going to church. His positions contradicted the official dogma. A conflict with the authorities was unavoidable.

At this time Friedrich II no longer occupied the throne of Prussia. He had been succeeded by his nephew, Friedrich Wilhelm II. The uncle had been a freethinking despot, a capable administrator, a sound general and a patron of the arts. The nephew, on the other hand, was a tepid, stupid man, inclining to mysticism. He was enthralled by magic and belonged to the secret order of Rosicrucians. Despotism was as strong as ever in Purssia, but enlightenment began to wither away. Friedrich's motto, "Reason as much as you like, but obey!", was replaced by the more customary, "Obey without reasoning!"

At first the relations between Kant and the new king bid well for the philosopher. Kant was *rector magnificus* when Friedrich Wilhelm II came to Königsberg to be crowned. As rector of the university Kant was invited to the royal castle, where he greeted the king in the name of professors and students; and the king, in return, paid him a compliment. (Kant declined to attend the thanksgiving service, pleading ill health.)

In his second term of office (1788) Kant opened the celebration of the king's birthday with the following words:

> The day which gave to the world our most beloved monarch is indeed a festive day for our university, for all of his true subjects, and for Europe herself in so far as it loves the peace founded upon justice and humanitarian concerns and made secure by power. Our university, ever mindful of its duties, will give today many proofs of our respectful and grateful sentiments to our most gracious monarch. Your excellencies have been pleased to join us in expressing our very best wishes on this day, and to grace the festivities with your august presence.[199]

The last words were addressed to the representatives of the authorities of the city. A draft of this text consisting of three sentences has been preserved: words are crossed out, then written in again, then replaced by different formulations. Kant did not find it easy to produce flattering formulations! But his position left him no choice, at least he left it brief.

He could allow himself to be ambiguous. The second edition of the *Critique of Pure Reason* came out in 1787. We must remember that the philosopher added a special section against any theoretical attempts at proving the immortality of the soul. That part is towards the end of the book, and what is at the beginning? There is a new foreword claiming that critical philosophy gave religion its first respectable foundations: "I must supersede knowledge in order to make place for faith." The "supersession" (*Aufhebung*) of knowledge was quickly understood by the obscurantists as a "setting aside" (*Beseitigung*), whereby there would be room for belief. "I must take knowledge into custody" was what the police spies heard. But the careful reader who read the *Critique* to the end had a different understanding: the philosopher had to "elevate" (*erheben*) knowledge in order to find a place for faith. We have already talked about the ambiguous use of the verb and its ironic function elsewhere.

The king supported Kant's nomination to the Academy of Sciences. Although no petition had been sent from Königsberg, Berlin considerably raised his salary: it was now seven hundred and twenty talers. In the fall of 1788 an able young man, Johann Gottfried Kiesewetter, was sent officially from Berlin to study critical philosophy at its source, so that he could lecture on it at the court. Kiesewetter spent a year at the Albertina. Upon his return to the capital he began to teach critical philosophy. Immediately it became obvious to him that the times had changed.

The revolution had broken out in France. At first there was general rejoicing in Germany. This did not arise solely due to the democratic tendencies of the lower orders: the Bourbons were considered enemies of the German nation, and their demise was greeted by the upper classes. The regimental band in Potsdam practiced the march of the sansculottes, fashion-conscious ladies of Berlin society wore sashes in the colours of the revolutionary government and the duchess of Gotha decorated her salon with bursts of the Parisian revolutionaries.

Then sobriety set in. There were fears that the fires of revolution could spread in Prussian territory. In Berlin they began to take the first precautionary measures. The Kantian von Zedlitz was replaced by Wöllner, "a prevaricating and intriguing priest, nothing more" — as the late Friedrich had characterised him. A new religious edict was issued in which the freedom of conscience was affirmed, but subjects were advised to "keep their opinions to themselves" and to refrain from broadcasting them. Then came the new censorship law, designed to limit the "excesses of the so-called enlighteners" and curb "the freedom of the press, which was becoming an outrage."

Kiesewetter was warned that he should say nothing against religion in his lectures and that he should not forget to point out that Kant's philosophy was not contrary to the spirit of Christianity. And indeed, Kiesewetter spoke only of this topic in his explanation of the *Critique of Practical Reason*. (And not gratuitously, either: the attention of the audience was drawn to an assiduously anxious young man who was busy trying to take down every word of the lecture, and who the next day failed to appear.) In his publications, too, Kiesewetter emphasized this same point. And he assured Kant: "I am convinced that it can at least be made clear that the foundations of your moral system are in full agreement with Christianity, perhaps even that if Christ heard you and understood you he would say that yes, that is precisely what I wanted to say about my Lord and my God, etc."[200] There were other men who thought that Kant was a troublemaker, a sceptic and a purveyor of doubt. There were rumours that Kant would be forbidden to publish. Also that someone had already asked the king to issue an order to that effect. Yet, Kiesewetter went on teaching Kantian philosophy to the ladies at court, and the education of the heir to the throne was also entrusted to him. The situation remained murky.

On grounds of the religious edict the "queue-haired" chaplain Johann Heinrich Schultz was brought to court. Kant had earlier reviewed his book on morality, and Friedrich II had saved him from prosecution. The brilliant orator Schultz had convinced his judges that his sermons, even though they denied the trinity of God, did not contradict Christ's teachings. The court ruled that Schultz was not a Lutheran but that he was a Christian preacher and set him free. The king ordered that the verdict be changed from innocent to guilty and imposed a heavy fine on the judges. Schultz was relieved of his duties.

The king was a plaything in the hands of clever favourites. They made use of his weakness (mistresses and the belief in ghosts) in their intrigues. The king was attracted to the countess Dönhoff; he was secretly divorced from the queen and married her. All of this did not spoil the field for another countess, of Lichtenau, who was considered to be his principal mistress.

The queen was declared insane because "she often danced on tables and chairs, and saw ghosts!"; thus Kiesewetter informed Kant. And things were not much better with her erstwhile husband who had already had various visions: the "spiritually and bodily weak" king "sat for hours and cried."[201]

The following amazing, although highly dubious, story gives a vivid illustration of what Friedrich Wilhelm II's contemporaries thought of him. Shortly before the battle of Valmy, Prussian troops took Verdun. In honour of this success and the imminent complete destruction of the sansculottes the king gave a ball. At the height of the festivities an unknown man approached the king, whispered in his ear the password of the Rosicrucians and asked the king to follow him. Ever obedient to the statutes of the order, His Highness complied forthwith. He was taken to a dark room, where there were no candles; only the dying embers in the fireplace threw ominous shadows upon dusty walls. The unknown man disappeared and the king wanted to return to his guests when a well-known voice stopped him in his tracks. (He knew the voice, there was no doubt about that, he had heard it a thousand times in the quiet of Sanssouci, in marches and on parades.) In the darkness he saw the known stooping figure — the sharp profile, the quick movements, the familiar coat and the cane. The ghost of his late uncle spoke to the king of treason, Prussia had been drawn into a dangerous adventure, take not a step farther, return immediately to your frontiers.

What followed in the days after the vision no one could understand. Prussian troops stopped where they were; and after Valmy, as there was no need to seek other decisive engagements, a general retreat began. Contemporaries and historians tried to understand it, and even Bonaparte could not find an explanation for the behaviour of the Prussians.

Beaumarchais broadcast the story about old Fritz's ghost. The dramatist had a friend, the actor Fleury, who was famous for his brilliant portrayal of Friedrich II; he could do a perfect imitation of the king's voice and bearing, and had somehow got hold of the monarch's coat and hat. In the turbulent September days of 1792, at the onset of the invasion, Fleury fled from Paris to Verdun. We are, however, interested in only one thing: the whole of Europe was talking about Friedrich Wilhelm II's vision.

Kant found out about it from Kiesewetter. He also knew that this was a disease of the times. Kant compared the "growing inclination towards fanaticism" to an epidemic of influenza ("Russian catarrh"). The age of the Enlightenment is not an enlightened age. Men are half-educated. They read much, know and understand little. And pass judgments about everything with an unbelievable ease! The visionary wants to know the source of "animal magnetism" (as hypnosis was called in the eighteenth century), the man of learning, naturally, refrains from proferring an answer, and the play of fantasy begins.

> There is only one cure for this conceit: to magnetise and disorganise the animal magnetiser for as long as it pleases him and the gullible multitude; the police, however, must see to it that there is no morally harmful behaviour; follow the procedures of science in order to test the properties of the objects of our senses and come to know them by means of observation and experimentation. Long-winded refutations are beneath the dignity of reason and superfluous here: a disdainful silence is a more appropriate way of dealing with such nonsense: such events are of short duration in the world of morality, they quickly yield place to other forms of stupidity.[202]

A quarter of a century earlier Kant had ridiculed Swedenborg, the seer of ghosts; now Count Cagliostro was all the rage. The Italian adventurer, Giuseppe Balsamo, who had taken the title and the name of

Count Cagliostro, criss-crossed almost all of Europe and made fools of high-society innocents by representing himself as an alchemist, a seer, and a magician and taking in large sums of money. In the eighties he appeared in Mitau, then went on to Petersburg and Warsaw. Catherine II herself wrote two comedies about him. Exposés began to appear in the German press. Kant's pupil Borowski wrote a book against Cagliostro. He also wrote to his teacher and begged him for his opinion of visionaries and seers. Kant answered him in a letter intended for publication. Borowski published it as an addendum to his work.

But something else had been expected from Kant for a long time: a book on the subject of religion. The sanctimonious accused him of lack of faith; his pupils, on the other hand, proved the orthodoxy of critical philosophy. Kant knew that sooner or later he would have to take a stand. Before he had completed his treatise he would read a work from another hand, which sought, however, to develop his thought further. The work was entitled *Towards a Critique of All Revelation*, and its author was Fichte.

It is necessary to say a few things about the founder of classical German idealism. Johann Gottlieb Fichte was born in 1762 (like Kant, the son of an artisan). The boy was endowed with a phenomenal memory — when he was nine he could repeat the Sunday sermon word for word. Another characteristic manifested itself very early — an uncompromising rectitude, a belief in his mission. "I am a priest of the truth; I am in its service; I have undertaken to do everything for it, to take risks for it and to suffer for it. If I suffer hate and persecution for its sake, even if I die in its service — what special act would I be performing, what further service than that which I absolutely must do?"[203]

His youth was spent in difficult circumstances. After his university studies he wandered from one house to another, teaching the children of the well-to-do. He was twenty-eight years old when a student asked him to help him in his studies of Kant's philosophy. Fichte's knowledge of Kant was hazy: he only knew that no one could understand him. Fichte therefore sat down with the *Critiques* and immediately saw the truth in them. Now he had only one burning desire: to come to know Kant as soon as possible and to complete his education under his guidance.

In the summer of 1791 Fichte came to Königsberg. As soon as he was able to he sought Kant out. His expectations were too high to allow fulfillment. Fichte met a tired old man, sunk in thought, indifferent of his guest and unable to sense his enthusiasm. Fichte found Kant's lectures disappointing; it was not possible to establish a closer relationship with the teacher. Then the young man took a decisive step. He cut classes for more than a month, worked feverishly, and addressed himself in a letter to Kant: "I came to Königsberg to become a little better acquainted with the man whom the whole of Europe holds in the highest esteem, but whom no one in Europe loves as much as I do. I presented myself to you. Only later I understood that it was impudent to expect to make the acquaintance of such a man simply out of the blue, as it were. I ought to have had letters of recommendation. I acknowledge only such letters as I write myself. Here, then, is my letter."[204] To the letter Fichte appended the lengthy treatise he had just completed after thirty-five days of strenuous work — *Towards a Critique of All Revelation.*

Kant was no longer surprised by spirited letters. Shortly before he had received another, even more extravagant: "Oh, great Kant. I come to you as a believer comes to his god to ask for help, for comfort, or for some knowledge of death; your reasons for an afterlife, given in your works, were enough for me, therefore I come to you seeking refuge." A certain Maria von Herbert had written from Austria. She had just had an unhappy love affair and held on to life only by reading Kant's works in ethics; the philosopher gave the letter to Pastor Borowski to answer. (Later he wrote himself — a lengthy, encouraging letter.)

But Fichte's manuscript did excite Kant's interest, the more so as it dealt with an issue on which he himself was working. Having read a few pages Kant realised that he had to do here with an extraordinary man. Fichte was invited to call, he came again to the Prinzessinstrasse and left disappointed again. Although he was received warmly, the conversation simply would not start. Fichte showered his host with questions, but the latter referred him to the *Critique of Pure Reason* and . . . to the court preacher Schultz. The third visit finally satisfied the student. Only now had he seen in Kant the features worthy of the great mind that also dominated in his writings — thus Fichte wrote in his diary. This time Fichte was invited to dinner: we already know that on such occasions Kant revealed his sharp mind and wit to their fullest.

Meanwhile, the devoted follower had fallen into worse straits. Money was running out, and there was no hope of employment in

Königsberg. Fichte took another decisive step. He begged Kant for help: "I have only two ducats left, and they are, properly speaking, no longer mine, for I owe them for my lodgings and the like. It seems that there are no means for my salvation if I cannot find someone who would be willing to lend to me, an unknown, a sum of money sufficient to pay for my trip back, until such time as I can return it, i.e., until the Easter of next year, and upon my word of honour. I know no one to whom I could make such a proposition without fear of ridicule, except you, oh, worthy man." Then Fichte went on for several pages about the spiritual torment connected with the request. In conclusion he said:

> I am, more by temperament and by experience than in principle, indifferent to what is beyond my powers. This is not the first time I find myself in a situation to which I see no resolution; but it would be the first time that I would surrender to circumstances. In such cases I usually only feel curiosity regarding their resolution. I simply employ the means that appear to be best suited and calmly await results. In this instance I can do that with greater reassurance because I have put myself into the hands of a wise and good man. But in another respect I send this letter with a good deal of trepidation. Whatever its outcome, I have forfeited some of the joyous sentiment that binds me to you. If the answer is yes, I can ultimately recoup some of what I have lost; if it is no, it seems to me I will have lost all.[205]

Kant's resolution was dialectical: neither affirmative, nor negative. Kant never gave alms in the street but was not averse to giving significant sums of money to his pastor for distribution to the needy. When his pupil Plessing was in financial straits (he had to leave Königsberg in a hurry in order to avoid being brought before the court for nonpayment of alimony) Kant unhesitatingly gave him thirty talers. (Plessing returned the loan nine years later — with the then customary five percent interest for each year.) Fichte got not a penny from Kant; but he did get something much better — a teaching post in a rich family, and an "introduction" to the life of great philosophy. Kant suggested that *Towards a Critique of All Revelation* be published, found a publisher, and arranged for an immediate payment of the honorarium.

The book attracted attention. It was published anonymously: although it was possible upon careful reading to distinguish in meaning and in style from Kant's works, its authorship was soon ascribed to the philosopher from Königsberg. He had for some time now been expected to produce a treatise on the philosophy of religion. It is also possible that the publisher brought out the book anonymously (against Fichte's wishes) so that it may be attributed to Kant. The little trick had a mighty effect. Kant had to publish a disclaimer and name the new name. Fichte became known.

While *Towards a Critique of All Revelation* was being prepared for publication, difficulties arose with censorship — the new times were making themselves felt. But Fichte was lucky: the censor was replaced and the book came out. Kant encountered more serious difficulties with the publication of his work on religion. The first part of the treatise, entitled *On the Radical Evil in Human Nature*, received the censorial imprimatur in the beginning of 1792 and was published in Biester's *Berlin Monthly*. Then the rules on censorship became more strict. A new royal order referred to the sad example of a great state where excessive freethinking had brought about the destruction of governmental institutions.

The state in question was France in the grips of the great revolution. The French king was the prisoner of the people and was awaiting trial. The influence of the clergy was waning, in the wake of dechristianization congregations were breaking up and churches were being destroyed. The fears of the Prussian king and his advisers that the destruction might spread to their own land can easily be understood.

When Biester received the second of Kant's treatises *On the Struggle between the Good and Evil Principles* he suggested that it bypass the censor (it was possible to do that). But Kant wanted no scandal concerning a trifling matter and therefore insisted on the compliance with the letter of the law. The treatise was banned. The same censor who had passed the first treatise refused to pass the second. An appeal to the king was unsuccessful.

Kant had already completed the whole work, i.e., all of its four parts. Thus it was now worth the risk of a scandal for the philosopher. It was not difficult to deceive vigilant censors: universities also had the right to authorize the publication of scientific literature. Kant submitted the manuscript to the School of Theology at the Albertina and obtained their answer: in so far as the treatise was philosophical in nature, they thought he ought to submit it to the appropriate school. Kant then sent *Religion within the Limits of Reason Alone* to liberal

Jena, where the dean of the School of Philosophy without hesitation gave the manuscript his *vidit*. The book came out in the spring of 1793 and caused no upheaval.

Instead of the banned treatise, Kant sent another to *The Berlin Monthly — On the Old Saying: It May Be Correct in Theory But It Does Not Work in Practice*. It is concerned with morality and right. The latter was becoming of ever more interest to Kant, as it was here that he sought the answer to the question of what man was to hope for. The Prussian legal system did not indeed give rise to optimism, but Kant kept his arguments general. He condemned despotism. One cannot force oneself to be happy after the fashion that others decree. Everyone has the right to seek his happiness in the way that he sees fit (as long as he does not prevent others from obtaining the same goal). The greatest despotism is a paternalistic government, where the subjects are treated as adolescents incapable of knowing what is right for them and what is wrong (the head of the state makes those decisions for them). The government should not be fatherly, but fatherlandish, bringing together all law-abiding men and women.

The essay *On the Influence of the Moon on Weather* followed. It, too, contained an ironic barb at the rulers. Science, Kant says, is a little like catechism. The older we grow the less we understand of it; actually, we should have to start school anew, if we could only be sure of there being someone there who understands things better than we do.

The treatise was written in April 1794 — shortly before his seventieth birthday. There were no birthday celebrations. Kant was glad that perhaps it had been forgotten. From day to day he expected the government to take strong repressive measures. "Life is short, especially what is left after seventy years have been lived; perhaps a corner of this earth can be found where one can end one's days in peace and tranquility. — If you tell me something that is not a secret but has not yet reached this place of ours, I would be glad." This request he sent to Biester, the publisher of *The Berlin Monthly*. When he sent Biester *The End of All Things* he asked him to publish it as soon as possible, "before there is an end to your and my literary activity."[206] Both of them had been threatened for some time. Nevertheless, one of them wrote and the other published a treatise which is a masterpiece of philosophical irony. But that was really the last straw; the patience of the government was exhausted.

Now, what in *The End of All Things* could have provoked such great displeasure? Kant treated the Bible without respect. He poked fun at the idea of the Last Judgment and at other church dogmas. "If by the end of all things we mean the end of the world in the form in which it now exists, namely, that the stars will fall from the firmament, that heavens themselves will collapse (or scatter like the leaves of a book whose binding has let go), that fires will consume all and a new heaven and a new earth will be created to bear the blessed, and a hell to receive the sinners, then such a day of judgment cannot be the last day for it will be followed by other days. The very idea of the end of all things stems from the contemplation of the moral arrangement of the world, not of its physical nature." In other words, what is at issue is the ultimate purpose of man's being. If that purpose is not attainable for rational creatures, then creation itself appears to have been purposeless: it is "like a play that has no conclusion and no rational intention that can be recognized."

In Kant's ironic view the end of all things can have three forms: 1. a natural form, which corresponds to the moral aims of divine wisdom; 2. a supernatural form, as the effect of causes which are beyond our comprehension; and 3. an unnatural form, which we bring about ourselves "by our misunderstanding of ultimate purpose." In saying this Kant was apparently referring to the unnatural character of the repressive measures which Wöllner's ministry had taken to strengthen the position of religion. What was to be done to strengthen religion? The best thing to do, according to Kant, was not to interfere with the natural course of events.

> I am so profoundly conscious of my inability to begin a new and hopeful enquiry that I only want to offer a piece of advice which does not require great perspicacity: leave things as they are, as they have arranged themselves in the course of almost a generation. Of course, men of great and enterprising spirit cannot be satisfied with this advice: so may I be permitted to call attention not to what they intend to do, but what they are likely to undo, so that they may not act against their own (even the very best) intentions.[207]

Kant's irony is tinted with melancholy. The reading of the treatise will afford both "sorrow and laughter," according to the author in a letter to Biester on 10 April, 1794.

In the text we find a clear and fairly daring attack on French freethinking. In his day Voltaire had said ironically: amid Moses' commandments the instructions for the construction of latrines were not forgotten. In the spirit of Voltaire, Kant parodies the biblical legend about the first appearance of the denizens of the earth: they are, so to speak, the result of a purge. The world of the earth, he writes, can be compared to "a sewer, to which all the dirt from other worlds is banished. This very original thought occurred to a Persian wit who moved paradise, the habitat of the first human couple, upwards into the heavens. There grew many trees, laden with beautiful fruits, the remnants of which imperceptibly evaporated after they had had their fill. There was one tree in that garden that bore a peculiar fruit — one very delicious but which did not evaporate. As our forbearers sated themselves with its fruit, in spite of the prohibition to do so, it was necessary, in order not to dirty the heavens, to take the advice of an angel who pointed at the earth and said: 'That is the lavatory of the universe.' He took them there so that they could attend to their needs and returned to heaven, leaving them on the earth. That is how first men came to be on the earth."[208] One can understand the fury of the authorities.

Rumours began to circulate that Kant was soon going to be called to account. It was said that he would be given the choice of either renouncing his views or leaving the university. That was how they had treated Kant's colleague Professor Hasse, who had been accused of freethinking. Hasse distanced himself from his earlier views and repented; now he was the subject of ridicule. Kant's resignation seemed unavoidable.

A letter came from far away Braunschweig, from Campe, with whom Kant had once corresponded. The former pedagogue of the Philanthropin did not doubt for a moment whether Kant would ever renounce his convictions and offered him his support: "Please consider yourself the master of everything I own; you would give me and those who are with me great joy if you moved into my rather roomy house, which would from then on be yours, and assumed the place of the head of my small family."[209]

Kant, deeply moved, answered Campe that he should pay no attention to the rumours. "The commandant of the city has given me no orders to renounce my views." But Campe was quite right in his assessment of how Kant would have acted if he had been presented with such a request: this could not have happened, however, because he had broken no laws. And even in the worst case, he had the means to live out the rest of his life without having to become a burden to anyone else, no matter how generously assistance was offered.

In July 1794 the Russian Academy of Sciences elected him to its membership.[210] Fourteen foreign scholars were elected to the Academy at the meeting on 28 July. Kant was nominated by I.I. Georgi, the geographer, who praised not only the *Critique of Pure Reason* but also the *Physical Geography*, which had not yet been published and which was known only as a series of lectures. In Russia Kant was not only known to the academicians. Ludwig Mellmann from Göttingen introduced the ideas of critical philosophy in Moscow: hc taught them in the nineties, first at the university *gymnasium* and then at the university itself. In 1791 N. Karamzin published the "Letters of a Russian Traveller" in *The Moscow Journal*; the letters contained an account of his visit to the famous philosopher.

In the same year a philosophical novel, *The Life of Faust, His Deeds and Road to Damnation* appeared in St. Petersburg. The novel shared not only the subject matter of Goethe's masterpiece (taken from a medieval legend), but also the Kantian understanding of human life and tribulations. The epilogue of the book contained an ironic wish that the philosophers of Germany might defeat their great adversary, the all-destructive Kant, so that metaphysical nonsense might resound from their lecterns for ever and ever. The author of the anonymously published work was the erstwhile leader of *Storm and Stress*, Friedrich Maximilian Klinger, now an officer in the Russian army with the name of Fyodor Ivanovich Klinger (he died a thoroughgoing Russian with the rank of a lieutenant general). Klinger's Faust is an enemy of feudalism and clericalism and filled with the consciousness of right. He invented the printing press, but observes with profound suspicion the dehumanising uses to which the sciences are put. In his expeditions through the world he comes across a group of scholars thirsting for knowledge who are studying the anatomy of a man who has been skinned alive.

In the summer of 1794 the Russian officer Woldemar Ungern-Sternberg visited Kant. Then he published a booklet in St. Petersburg, *An Address to Russia*, with an ode to Catherine II and a dedication to the great men of the world — Count Rumyantsev-Zadunaysky, Prince Zubov, Field Marshal Suvorov, etc. The name of Professor Kant completed the list. The author called him a prince of philosophers, a Caesar of the wise, a great man without titles and decorations. He contended that Aristotle had already foreseen Kant's appearance, and that Empress Catherine had realized his ideas on the institutions of the bourgeois civil society. Catherine's laws made moral freedom possible, she taught the obedience to laws; huts and palaces were decorated with the same roses, all the citizens drank from the same cup. Kant did not answer Sternberg.

In his answer to Prince Beloselsky, Kant weighed every word carefully; their exchange of letters is one of the most interesting yet almost totally forgotten episodes in the history of Kant's lifelong relationship with the culture of Russia. We shall treat of it separately at the end of the book.

Kant was elected to the Petersburg Academy of Sciences in July 1794. In September he received a letter from the corresponding secretary Euler (the son of the famous mathematician): Euler told him of the election and promised to send the diploma at the first "safe opportunity," as soon as the imperial seal had been affixed. Life was easy-going in those days, and the postal services were irregular. But the diploma finally safely reached the recipient. Here is the text:

Serenissimae et Potentissimae Imperatricis
CATHARINAE SECUNDAE AUGUSTAE
totius Russiae Autocratoris
iussu

Ego CATHARINA PRINCEPS DASCHKAW
ab intimis AUGUSTAE Cubiculis Ordinis St. Catharinae Eques
Academie Scientiarum Director

pro potestate ab AUGUSTA mihi concessa
Virum celeberrimum Suisque titulis condecorandum
Immanuelem Kant, Professorem Philosophiae Regiomonti
ob praeclara ejus in Scientia merita
Communi totius Academiae Petropolitanae suffragio
Collegam hujus Societatis externum
solemni hoc Diplomate declaro
eumque honore, privilegiis et beneficiis Academicorum
ordini concessis rite orno
Petropoli A.C. MDCCXIV, die XXVII Julii
(Signed) Princeps Daschkawie
Johannes Albertus Euler
Academiae Conventus Secretarius perpetuus
Consil. Aul. et Ordinis St. Vladimeri Eques.

It is quite possible that Kant read the encouraging words from St. Petersburg at the same time as the censure from Berlin. For the storm had finally broken out. The Prussian government had been thinking hard and long about how to punish the world-renowned scholar without losing face. Finally they found the way: in October 1794 Kant received a royal order, but no one — except the philosopher — knew about it. The special rescript was not made public, it was sent as a private letter to Kant.

> Our highest person has for some time observed with greatest displeasure: how you use your philosophy to distort and bring low some of the principal and fundamental teachings of the Holy Scripture and of Christianity; how you have done that particularly in your book *Religion within the Limits of Reason Alone* and in other, shorter treatises. We expected better of you, since you must understand how irresponsibly you discharge your duty as the teacher of the young and act contrary to our well-known goals for the fatherland (after Kant had made fun of the idea of a "fatherly" government it was risky to use the term in an official document!). We expect a speedy and conscientious explanation and hope that in order to escape our highest displeasure you will in the future refrain from such behaviour and will rather endeavour to use your talent and your influence to realize our intentions for our fatherland more and more; should you choose to do otherwise and persevere in your conduct, you will call upon yourself unpleasant consequences.[211]

Wollner signed the despatch in the name of the king.

To renounce his views was against Kant's rules but he was not strong enough to resist. He outlined the only tactics open to him on a piece of paper that happened to be at hand: "A renunciation and abjuration of one's inner convictions is despicable and can be demanded of no one: but in a case like this silence is the duty of the subject, and although everything one says must be true, it is not therefore also only's duty to publicly proclaim all of the truth."[212] Thus the categorical imperative assumed a concrete historical form.

Kant was asked to reply without delay, and he replied without delay. He observed all the dignified formulae that a subject uses when addressing his monarch but he did not repent; on the contrary, he refuted point by point all the charges that had been made against him. First, as "teacher of the young" he had not gone beyond the limits of philosophical enquiry into religion in his lectures; an inspection of his students' notes would bear out this contention. Second, as the author of *Religion within the Limits of Reason Alone* he did not go against the "intentions of the august sovereign," because the latter was interested in the good of the religion of the land, and the book was intended for the use of the students of theology and philosophy so that they might find out how best to bring religion into the hearts of men, something about which the people have absolutely no idea. Third, there were no attempts to degrade Christianity in the book mentioned, because it contained no judgments regarding the revelations of existent religions; only the religion of reason is analyzed there as being the highest prerequisite of every true religion. Kant wanted his accusers to show him a single instance where he had permitted himself to degrade Christianity, had denied it as a revealed doctrine, or called it unnecessary. Fourth, Kant insisted that he had always shown true respect for Christianity, in that he taught that the Bible was the best means for establishing and maintaining a moral national religion. And fifth,

> Just as I have always demanded that other proponents of religions be conscientiously scrupulous, that they make no claims about it other than those of which they themselves are thoroughly convinced, and that they not make others believe that of which they were not absolutely certain, so too in my works on religion I never lost sight of this judge in me and have imagined him at my side in writing my works, in order not only to prevent error dangerous to the soul, but to avoid any carelessness in expression which might prove upsetting; this is why now, in my seventy-first year, when the thought comes inevitably that I may shortly have to answer to the judge of all creation who knows what is in our hearts, I can in good conscience give this account to the authorities.[213]

And finally, sixth, in order to avoid giving cause for charges that he was distorting and degrading religion, the philosopher promised, "as the faithful subject of His Royal Majesty," that he would refrain from any further public pronouncements about religion.

Kant's answer was worthy of the master of irony. "As the most faithful subject of His Royal Majesty" — the respectful formulation was ambiguous: after the death of Friedrich Wilhelm, Kant declared that he was now released from the obligation undertaken because he was now the subject of another "Majesty." In *Strife among the Faculties* Kant returned to the explication of the Bible and published his correspondence with the king in the foreword to the work.

The *Strife among the Faculties* is, in Kant's own words, "actually mere journalism"[213] — polemical and ironic. That is especially true of the first part, written in 1794 and under the impression of governmental censure. The three "higher" faculties — the faculties of theology, law, and medicine — do not rely on reason, but on statutes. The theologians appeal to the Bible, the jurists to the law of the land, and even the physicians do not attend to the physics of the human body in their therapies but to medical textbooks. (In general, however, the medical faculty is much freer than the first two and stands "quite close to the philosophical faculty.")

The strife among the faculties has to do with their influence on the populus. The people want to be lead, and in the vocabulary of the demagogues that means: deceived. Kant ridicules the philistines: "What you philosophers blab about I have known for a long time; but I come to you as men of learning to learn how I can obtain a ticket to heaven at the last minute after a wicked life; how can I win a judgment even when I am in the wrong, and how can I remain healthy and live long even if I have squandered the powers of my body in lustful and riotous living."[214]

The higher faculties are the right wing in the parliament of learning, the ruling party; the philosophical faculty is the left wing, a kind of party of the opposition, and in virtue thereof quite an indispensable institution for the state. Without its rigorous testing and objections the government will not have a very clear idea of what is harmful and what is beneficial to it. The philosophers have been called upon to control the three higher faculties; otherwise they are a thoroughly modest lot: "one can agree with the proud claim of the theological faculty that the philosophical faculty is its handmaiden (although the question still remains of whether this servant carries the torch in front of the gracious lady, or the train of her dress behind her), as long as they do not close down the philosophical faculty or shut its mouth." Kant believes in progress and hopes that with the passage of time the situation will change: "the last will be the first, the lower faculty will be the higher," not in the sense of a sovereign authority (philosophers are not seekers of power) but in the sense of "counselors to the mighty (the government) who in the freedom of the philosophical faculty and the insight ensuing from it will prove a better means of achieving the goals of government than its own absolute authority."[215]

In the *Strife among the Faculties* Kant obstinately repeats that church belief cannot be identified with religion. "If I accept faith . . . as a principle, then such faith is . . . not part of religion."[216] Faith in the teachings of the Bible is in itself not a virtue but a lack of faith, and doubt in itself is not a sin. ". . . in religion everything depends on deeds."[217] Theoretical reason is the crown of knowledge, practical reason is not faith, but action.

In the beginning was the deed. That is how Goethe's Faust understood the gospel according to John. Action is the principle of being, as a new interpretation has it. Kant could have made it more precise: in moral action lies the principle of man's being. The "strife between the pnilosophical and theological faculties" concludes a reflection regarding the specific feature of man. The understanding does not separate him from the animal, because instinct is also a "kind of understanding." Only morality makes man man. If he loses it, he turns into an animal.

After the *Critique of Judgment* the problem of man is at the centre of Kant's concern: it is no longer latent, but out in the open. To the three earlier fundamental questions of philosophy he now adds a fourth — the most important: the question of man. Here is a document from the middle of the nineties: "The field of philosophy in this cosmopolitan sense can be summed up in the following questions:

1. What can I know?
2. What ought I to do?
3. What can I hope for?
4. What is man?

Metaphysics provides an answer to the first question, morality the second, religion the third, and anthropology the fourth. Ultimately one can reduce all of them to anthropology, because the first three questions are related to the last."[218]

Man is born to society, they said in Kant's day. As time went on Kant thought about this point more and more. The acknowledgment of the formative influence of culture (and especially art) as a universal principle was not the whole answer to the question of man for Kant. He had uncovered contradictions precisely in cultural development which could not be eliminated by either the categorical imperative or the artistic intuition. The human being in itself can do nothing here. What, then, can he hope for? Kant brings sociology to bear on the religious and ethical solutions of the problem. Man can put his hope in others, in the society as a whole, in its social and legal institutions. The philosophy of right is a new frontier in Kant's spiritual development, one he came to when he was an old man. Kant's philosophy is optimistic. He not only loves man, he believes in him, he puts his hope in man's reason, his will, his nobility. Faith, hope, and love — this triad is as important to Kant as the already known triad of the good, the true, and the beautiful.

What can be more consoling to the man about to leave this world than the knowledge that he has not lived in vain.
— KANT

CHAPTER SEVEN

Perpetual Peace

The age of the Enlightenment ended in wars. The reign of reason, once announced by the ideological leaders of the third estate, was transformed into a bloody oppression and struggle for political and economic hegemony.

Revolutionary France was not prepared to wait for the onslaught of inimical forces and itself decided to attack. The National Assembly declared war on the "tyrants of Europe." The war was unpopular in Prussia. Goethe, who took part in the campaign, did not hide his contempt for the French émigrées who dreamt of a restoration of the monarhy. After the disaster at Valmy he made the famous pronouncement: "Here and now begins a new era in the history of the world."

The French now went on the offensive. Their troops forged ahead into German lands. Mainz and Frankfurt fell to them. A revolutionary convention was established in Mainz which announced the city's annexation to France (among the leaders of the Mainz group were

Kant's opponent Forster and Kant's follower Dorsch). A large tribute was levied upon Frankfurt: in Paris they were of the opinion that the war should support itself, and if possible the whole French nation. The French could not hold Frankfurt, and in the summer of 1793 they surrendered Mainz. The war continued, with changing results, until the April of 1795.

The news of events in western Europe reached Königsberg very late. Kant now read the papers not in the evening, during his leisure time, but as soon as they came out. Politics interested him more and more. The peace of Basel ended the fighting between France and Prussia, but it did not end the state of war: it was clear that new conflicts were in the offing. The philosopher drew up an account of the situation in the famous treatise *Perpetual Peace*, in which theoretical thoroughness is organically bound up with political realities and expressed in a nuanced ironic form.

The title itself, *Perpetual Peace*, must have sounded ambiguous to the German ear: at first it sounds like the stereotypical subtitle of a scientific treatise; then it reminds one of a no less stereotypical signboard on an inn. Kant did not fail to make use of this ambiguity. "Whether this satirical inscription on a Dutch innkeeper's sign, upon which a burial ground was painted, refers to mankind in general, or to the rulers of states in particular, who are insatiable of war, or merely to the philosophers who dream this sweet dream, it is not for us to decide."[219] So begins the treatise on peace.

The very term, "perpetual peace," is ambiguous in that it reveals two possible solutions to humanity: either the end of all hostilities through an international agreement, or the perpetual peace of the gigantic cemetery of mankind after an annihilating war. Everybody could dream about the first possibility; in the eighteenth century the second possibility revealed itself only to a man who was accustomed to thinking everything out to the end. Many were writing about the first possibility in those days, but only Kant referred to the second.

But Kant is otherwise an optimist and works tirelessly for an understanding among the nations. He develops his treatise as a contract, a parody of similar diplomatic documents. First come the preliminary articles, then the definitive articles; there is even a secret clause. The preliminary articles formulate the conditions which ought to clear the way for normal relationships among the states. 1. No treaty of peace shall be held valid in which there is a tacitly reserved matter for a future war. 2. No independent states, large or small, shall come under the dominion of another state by inheritance, exchange, purchase, or donation. 3. Standing armies shall in time be totally abolished. 4. National debts shall not be contracted with respect to external disputes of the state. 5. No state shall interfere by force in the constitution or government of another state. 6. No state shall, during war, permit such acts of hostility which would make mutual confidence in subsequent peace impossible: such are the employment of assassins, poisoners, the breach of capitulation, and incitement to treason in the warring state.

In the "definitive" articles Kant treats the maintenance of established peace. The civil constitution of every state should be republican. (A republic, in Kant's view, is not incompatible with the monarchical structure: the necessary requirements are the rule of law, openness, and the distribution of powers; Kant thought that Friedrich II ruled in a republican fashion.) The second definitive article for perpetual peace sets the grounds for a universal international law: it must develop a voluntary league of nations with a constitution like that of an individual republic, i.e., one in which the rights of every member state are guaranteed. The league of nations, a "federation of free states," is not a world state; Kant openly advocates the maintenance of national sovereignty. The third definitive article limits the law of "world citizenship" to conditions of universal hospitality in a foreign land. Each man must be able to visit every corner of the earth without becoming subject to acts of enmity. Each nation has the right to its own territory: foreign newcomers should not threaten it with oppression. Kant is an enemy of colonialism.

He was of course not the first to understand the misery of war. Even in ancient times men had thought about ways to eliminate bloodshed in human society. Erasmus of Rotterdam called upon the monarchs to desist from war. In the seventeenth century the idea was born of establishing perpetual peace by means of treaties among all states. St.-Pierre and Rousseau were dedicated proponents of the idea. The new argument which Kant brought into the debate was the grounding of the inevitable necessity that a perpetual peace be established upon the earth. War as a form of international behaviour will not be set aside by the arbitrary will of monarchs, but by the force of historical necessity. States will ultimately create a world union, just as men once created the first state. In both instances some rights (in particular the right to self-determination) must be given up; this, however, is very compatible with the maintenance of independence. Kant had already touched on this in 1784, in the treatise on universal history. He had maintained that view and reiterated it incessantly. Kant was not seeking to convince monarchs; he did not scourge them for their worldliness, as his predecessors had done. He insisted on the inescapable necessity for peace and ridiculed those who did not take note of the force of the times.

A secret article crowns the treaty on perpetual peace. What does it contain? It is only a jest: "The opinions of philosophers on the conditions of the possibility of public peace shall be consulted by the states armed for war." All the stipulations of the article are ironic and mocking. Here is the passage dealing with lawyers: "The lawyer, who has made not only the scales of right but also the sword of justice his symbol, generally uses the latter not to keep back all foreign influences from the former, but to throw into it the scale if it does not sink the way he wishes it to." And here he takes on the philosophers who in accord with the Platonic utopia aspire to power: "That kings should philosophise or philosophers become kings is not to be expected. Nor is it to be wished, since the possession of power inevitably corrupts the untrammeled judgment of reason. But kings or king-like peoples who rule themselves under laws of equality should not suffer the class of philosophers to disappear or to be silent, but should let them speak openly, because that is indispensable to the enlightenment of the business of government . . ."[220]

And what kinds of advice does Kant offer to the practitioner of politics? 1. *Fac et excusa.* Seize every favourable opportunity for usurping the right of the state over its own people or over a neighbouring people; the justification will be easier and more elegant *after* the deed, and the power can be more easily glossed over than if one has first to wait upon the consideration of convincing arguments and to meet them with counterarguments. 2. *Si fecisti, nega.* When you have commited an injustice—for instance, bringing your people to despair and hence to rebellion—deny that it was your fault; assert that it was done due to the obstinacy of your subjects. 3. *Divide et impera.* "That means, if there are certain privileged persons in your nation who have chosen you as their chief, set them at variance with one another and with the people: show the latter visions of greater freedom and all will soon depend on your untrammeled will. Or if it is a matter involving foreign states, it is a pretty safe means to sow discord among them so that under the guise of protection of the weak you can conquer them one after another." Certainly, Kant concludes, no one can be duped by these political maxims, for they are already universally known. Practitioners of these maxims are not ashamed of revealing them; they are ashamed only when they fail.

How does one bring politics in harmony with morality? There are two possibilities: either adapt morality to the interests of politics, or subordinate politics to morality. The first mode of conduct is the choice of the "political moralist," who begins where the "moral politician" leaves off (in that he "puts the cart before the horse"): he subordinates principles to his purposes. A genuine unity of morality and politics is possible only on the foundations of right. Public knowledge thereof is the guarantee that this is the case.

No other work of Kant's excited so lively and immediate a response. The first edition of *Perpetual Peace* was literally snatched from the hands of booksellers. The publisher, Nicolovius, brought out the second edition in the same year, 1795. A French translation was published simultaneously in Bern, but Kant was not happy with it. In 1796 Nicolovius brought out another German edition and a French translation, approved by Kant. A third translation was published in Paris, both as a book and in installments in the *Moniteur*. The semi-official newspaper declared: "The renowned Kant, who has caused a spiritual revolution in Germany equal to the revolution which crushed the Ancient Régime in France, has now lent the force of his name to the cause of the republican system of government."[221]

It is interesting that the correspondence between the ideas of critical philosophy and the spirit of the French Revolution was grasped so early. Heine felt the points of contact most keenly. (Marx called Kant's philosophy a German theory of the French Revolution.) Heine only erred in the choice of his equivalent characters. It was the moderate Sieyès, not the extremist Robespierre, for whom Kant would have been able to become enthusiastic. Sieyès sought to organize lectures on Kant's philosophy in Paris; the only reason the attempt failed was because there were no specialists who could do it. When Wilhelm von Humboldt came to Paris in 1798, Sieyès begged him to explain the fundamental ideas of critical philosophy to the French scholars. A five-hour colloquium was held to which famous "metaphysicians" (Cabanis, Destutt de Tracy, and others) in particular, were invited. Von Humboldt afterwards claimed that he had not been able to make himself understood to his audience, because the French were not disposed towards abstract thought. But there were other opinions on the matter: the lecturer was not entirely at ease with Kant's *Critique.*

There were rumours that Sieyès was going to send the new constitution to Kant for his approval; that Paris was going to ask Berlin to order the philosopher to France in order that he might help there with the improvement of the organization of the state. In 1797 a book was published (its author and the place of origin were unknown), entitled *Professor Kant's Answer to Abbé Sieyès*; the book contained a Christian utopia, and at the end of the book the author admitted that it was a forgery.

General Bonaparte showed a lively interest in everything unusual. Kant's philosophy was no exception. When the first consul was in Geneva, some unluckly scholar tried to convert him to Kantianism. Shortly after that, in Lausanne, Bonaparte asked one of the local scholars what the Swiss thought about Kant's philosophy. He received the answer: "General, we do not understand it at all," and gleefully said to his companion: "Well, Berthier, have you heard? Here they do not understand Kant either!"[222]

Finally Bonaparte found out that there was a prominent Kant expert in Paris, the former emigrant Charles de Villers: he was the author of a French exposition of the *Critique of Pure Reason* which had also been translated into German and published in Germany. The first consul summoned de Villers and told him to prepare a four-page summary of Kant's philosophy in four hours. De Villers produced a clear conspectus which failed to impress Bonaparte. When the future emperor concluded the concordat with the pope, he was pleased to say: "Priests are worth more than Cagliostro, Kant and all the German dreamers!"

Events in central Europe strengthened Kant's already growing interest in the problems of international law. Morality provides an inner law for human behaviour; in legal principles the inner conviction is combined with the external restraints. Finally a force is developed that regulates the life of society in so far as it strengthens morality and delivers man from the arbitrary exercise of another's will.

Law is formal. It is binding on all and admits of no exception. The smallest exception in the application of a law makes the law weak and useless. Kant had already said this about morality: it holds true in law as well. Naturally, we encounter difficulties and contradictions here too. Kant does not close his eyes to them and even points out two instances in which law is ambiguous.

The first case is the demand for justice (fairness). For instance — a man complains because he is paid with greatly devalued money at the conclusion of a contract, having entered the contract when the value was much higher: now each party has the stipulated amount, but the loss of value means that the buyer can buy less. That is unjust, but there is no legal remedy for the grieved party's claims. He can only appeal to fairness, the dumb godhead whose voice will not be heard. From the standpoint of justice this strict (narrow) right is the highest form of injustice. But there is no recourse. The law has to be observed. That is the axiom of the consciousness of the law.

The second case of ambiguous right pertains to right of necessity. It is said: "Need knows no commandment." Nevertheless, Kant says, there can be no need which can make injustice into action in compliance with the law. What holds true in the breach of moral norms, holds true here as well: if man is forced to break a law he must admit it and not represent evil as good, the breach of law as the obedience to it.

Kant's *Metaphysics of Morals* is a panegyric on the consciousness of the law. It came out in two separate parts, in January and August of 1797. The first part is dedicated to right, the second to morality. In the philosophy of right Kant finds the essentially completed answer to the question of "what can I hope for?" In addition to man's hope in himself Kant introduces the hope in society, in social institutions, and in duly constituted laws.

As the result Kant's moral doctrines undergo a change: they lose their rigorist nature. The thinker continues to be beset by "casuistic questions," to which it is impossible to give unequivocal answers. He becomes more tolerant, places fewer demands on man, finds more of a readiness to forgive man his transgressions. He speaks about the happiness of man as the ultimate purpose of the human kind, about love as the force which makes happiness possible. The second part of *Metaphysics of Morals* contains essential revisions of the positions in the *Critique of Practical Reason* and other works in the area of ethics.

Let us first take a look at the first part of that work. According to Kant, law is divided into private law and public law; the first pertains to the relations between private parties, the second to the relations between the individual and society, and between social groups. The major problem of the private law is property. Private property is the foundation of civil society, but not from the very beginning: "mine" and "thine" are a product of history. Only things can be objects of private property; man is purely its subject. Man must not take another as his property. It is true that there is a sphere of object-personal law, where men consider each other as things and use each other as commodities of exchange. This occurs in marriage, which Kant defines as "a union of two persons of opposite sexes for the purpose of potential possession of the other's sexual characteristics." Kant consistently underlines the equality of both parties in a marriage. Therefore it is not only man who can demand the return of his wife, if she leaves him, but also the other way around. Both have equal rights to pleasure. Here the views of the old bachelor Kant are considerably broader than the views of his married successor Fichte, who genuinely believed that only man should find pleasure in the act of sexual congress. Legal attributes of the citizen are freedom, equality, and independence. The first two are taken from the motto of the French revolution and are directed against all forms of feudal dependence, despotism, and class differences: all are equal before the law, and from that follows freedom and equality . . . The third device from the flag of the sansculottes was a mystical concept: brotherhood; Kant crowns his triad of legal attributes with a clearer concept: "civil independence." In Kant's view children, women, and servants are dependent: he therefore denies them the right of suffrage. But he does not deny them the right in general; he repeats stoutly that all are equal before the law. As pertains to the rights of suffrage we should remember that the Jacobin constitution of 1793 denied those rights to household servants. Such was the spirit of the time.

In order to eliminate despotism Kant insists on the strict division of powers. The idea is not new, but in the *Metaphysics of Morals* it is developed with extraordinary consistency and conviction. In every state there are three powers: the supreme or legislative power, the executive power which rules in obedience to the existent laws, and the judiciary, who are concerned with the observation of laws. Despotism arises when any one of these powers is not sufficiently independent of the other two. The opposite of despotism, in Kant's view, is the republic. If the principle of the division of powers is actualized in a constitutional monarchy, then we have a republic here too. Kant was called a royal Prussian republican. Monarchy (autocracy) is the simplest and therefore the most convenient system of government. It is always in danger of turning into despotism, but democracy is not safe from this danger either (when it becomes distorted, democracy turns into ochlocracy — the despotic rule of the mob). Kant does not attribute great significance to the form of government. The most important thing for him is that the land be ruled by laws, and not by men. Then none of the three powers is likely to overstep its limitations.

The legislature incorporates the unified will of the people. The legislator cannot simultaneously be the executive, because the first gives the laws and the second is subject to them. Neither the legislator, nor the executive can sit in judgment; they only appoint judges. The people rule themselves through representatives chosen in free elections. From Kant's pen flowed demands which are remarkably in harmony with the programme of the French revolution. They only differ from the latter in the choice of means.

Kant's thesis is clear:

> The duty of the people to suffer the abuses of supreme power, even those that may be considered insupportable, lies in the following: that their resistance to the supreme authority must never be thought of as anything but illegal, otherwise the whole legal structure would be destroyed. . . . The change in the (inadequate) constitution of the state, which from time to time becomes necessary, can only be effected by the sovereign himself by means of reforms, and not by the people by means of a revolution . . .[223]

This is clearly and definitely stated. Kant has been shaken by the development of events in France, and especially by the execution of Louis XVI at the hands of the Convention. The murder of a monarch during an uprising was not even the worst of it: "It is formal execution which horrifies the soul of one having ideas of human rights . . ."[223] That is the suicide of the state.

Now we must find the antithesis! And it is literally on the following page: "Once the revolution has succeeded and a new constitution has been established, the imbalance of the beginning does not absolve the subjects from the duty to submit to the new order of things as good citizens, and they cannot refuse obedience to the newly constituted authority." In the *Strife among the Faculties* Kant praises the French revolution: "The revolution of a gifted people which has been unfolding before our eyes may succeed or fail; it may be so filled with suffering and outrages that a sane man would not decide to undertake the experiment for the second time, even in the hope of a successful outcome, because of its cost — that revolution, I say, still excites a sympathy bordering on enthusiasm in the hearts of all observers . . ."[224] For it is the struggle for the violated rights of the people.

Oppression always prepares the way for uprising. Kant looks to the East, where the people of a gigantic empire are denied their rights; he asks: "Are we to see another revolution, this time undertaken by the Slavic peoples?"[225]

Kant is a sworn enemy of tyranny. He only fears that the use of violence in the struggle against it can shatter the sense of the law and thus bring about an even worse tyranny. The despot must be eliminated, but only by legal means. The people have "their inalienable rights against the head of the state . . . , yet these rights cannot be coercive." Well, what are these inalienable rights? First and foremost the right to criticism. "Freedom of expression — practiced within the limits of the respect and love for the social order in which man lives, through the liberal thought of subjects (and the self-imposed discipline of the pen, so that freedom may no be lost by abuse) — that is the sole safeguard of the rights of the people."[226]

One cannot indict the head of the state; one can only attempt to escape his power. Where and how, Kant does not explain, but the sense of his argument is clear: popular opinion can refuse to support the tyrant; placed in the condition of moral isolation and surrounded by an incipient uprising, the tyrant will have to attend to the voice of the people, respect existing laws or change them if they require improvement. The dissatisfied citizens must restrain themselves. Impatience has no place in this schema of things.

Among the proponents of Enlightenment, Kant's contemporaries, there were some who harboured anarchistic notions. Herder polemized strenuously against the state, calling it a machine which had to be destroyed from time to time. Kant sees that without the state there is nothing; he does not see it as a mechanism but as an organism, a living whole. Each part in such an organism should not be only a means, but also an end. A mechanical intervention in the life of this whole is not permissible. Universal progress is possible only through gradual perfection. The principal sign of such a progress is an improvement in the laws. A central passage in the *Strife among the*

Faculties is dedicated to grounding of the thesis that progress is also to be found in the area of legislative activity. We have already quoted the spirited evaluation of the French revolution, included in that work. Events of that kind, Kant says, are not forgotten in the history of the world; they reveal to mankind the inclination and the capacity for perfection; this is the triumph of the idea of law, in its essence not a phenomenon of the revolution but of an evolution in natural law.

(The solemn part is followed by an ironic one in which Kant makes fun of the idea of progress, i.e., makes fun of his own convictions. If in his exposition of religious conceptions Kant used irony to conceal, now it is put to a different use. The attention of the reader is not to be deflected, but drawn to the heart of the matter. Voltaire, among others, had used the same technique.)

Kant's statements on progress in the *Strife among the Faculties* are brought to an end with an anecdote. A physician comforted his patient by telling him that he discovered new symptoms of recovery each day. Once he praised his pulse, then his excreta, then assured him that perspiration was a sign of improvement. When the patient was asked how he felt, the poor wretch answered: "I am simply dying of recovery."

The joke was far ahead of its time. It appears to contain a warning against progress, which can also have lethal consequences. Did Kant foresee those consequences? Probably, because the joke is followed by another, which names the object of his worries: war. The section on the conflict with the school of law is crowned by a quotation from Hume: "When I now behold nations engaged in wars," Hume says, "I am reminded of two drunks fighting with cudgels in a china shop. It is not enough that it will take a long time for the injuries they have inflicted upon each other to heal, they will also have to pay the bill for the property they have destroyed."[227]

The idea of perpetual peace is the final link in Kant's philosophy. Wherever and whatever Kant wrote about society, his thought always and inevitably turned on the question of how to eliminate war.

The *Metaphysics of Morals* is no exception. In comparison with *Perpetual Peace*, however, it contains an essential correction. There universal peace was treated as a goal capable of practical achievement. Here, Kant is far more realistic: "Perpetual peace (the ultimate goal of international law) is, of course, an idea incapable of actualisation. But, the political principles designed to bring it about, namely the establishment of such international relations as would serve mankind's steady and ceaseless progress towards perpetual peace, are capable of actualisation. This task is not only a duty, but also a well founded right of man and states."[228]

The alternative to a universal peace secured by agreement is, as we recall, the eternal peace of the graveyard of mankind, an unnatural end to all things. No matter how utopian the idea of perpetual peace may appear, working towards it is an imperative of every foreign policy. It is an imperative of hope.

The *Metaphysics of Morals* came out in 1797. It was followed by the *Strife among the Faculties* — a work to which we have already referred three times: first in connection with the problem of improving laws (*The Strife between the Philosophical Faculty and the Faculty of Jurisprudence*); then in dealing with Kant's *contretemps* with Prussian censorship (*The Strife between the Philosophical and Theological Faculties*), and, finally, in the fourth chapter, in considering Kant's "system of health" (*The Strife between the Philosophical and Medical Faculties*). The strife among the faculties is a book written as a paean to reason. It appears that three self-contained essays, written at different times, are pulled together by a single thought: man's will and intellect are all-powerful, they can guide society on the road to progress, they can overcome prejudice and obscurantism and govern even the physical processes of the organism.

The last point was becoming a preoccupation with Kant as years went by. Physicians knew of Kant's (not at all amateurish) interest in medicine and often sought his advice. The renowned anatomist Sömmering sent him his manuscript *On the Organ of the Soul*. Kant's detailed answer was then published as an addendum to Sömmering's work. Kant repeated his long-held idea that mechanical-regulatory explanations are insufficient for an understanding of the activity of the organism and its nervous system. But there is something new in his answer — the striving to explain life in a natural fashion. Kant suggests that all the processes that entail the brain should be considered from the point of view of chemical reactions.

The famous physician Christoph Wilhelm Hufeland sent Kant his book *Macrobiotics, or the Art of Prolonging Life*. Kant answered with an essay "On the Power of the Mind to Master Its Feelings of Illness"; this article concludes the *Strife among the Faculties*.

Kant's power of mind had ruled over his body for a long time. But everything must come to an end. Kant was not ill, but his strength was failing. He taught less and less, ending his semester in February instead of April. He gave his last lecture (on logic) on 23 July 1796. For the next three semesters he announced courses with the proviso that they would be given health permitting. But his health no longer permitted, and the courses were cancelled.

In the course of his academic career Kant had taught two hundred sixty-eight courses; fifty-four on logic, forty-nine on metaphysics, forty-six on physical geography, twenty-eight on ethics, twenty-four on anthropology, twenty on theoretical physics, sixteen on mathematics, twelve on law, eleven on the encyclopaedia of philosophical sciences, four on pedagogy, two on mechanics, and one each on mineralogy and theology. Now his voice was to be heard from the lectern no more. Students wanted to honour their famous professor, but it was impossible to find any "round" date: Kant had begun teaching in 1755. And it was too early for any final farewell, because Kant did not yet want to be emeritus.

Then someone remembered that Kant's first work had come out in April 1747. So they decided to celebrate the fiftieth anniversary of the philosopher's literary activity in June 1797. The students organized a parade down to the house on Prinzessinstrasse. Orchestras played. A young man entered the room, congratulated Kant on the anniversary and assured him that no one would ever forget his lectures. In the street they were giving three cheers.

Kant was proud of the fact that in spite of his age he had retained his clarity of thought, stoutness of spirit, and an interest in life and his beloved work. He was no longer strong enough to lecture, but he could write. As before, he spent his time before dinner at his desk. He still attempted to take part in the life of the university. When the rector decided to exclude Kant from the university senate (the philosopher did not attend its meetings), he protested and got his way. In his letter to the rector Kant claimed that the principal work of the senate was accomplished by secret ballot and that one could cast one's vote at home as well, since it was deposited in a sealed ballot box: at home one is free of interference and has the time to think. From Berlin the order came to keep Kant in the senate as long as he wished to remain in it.

In the summer of 1797 he discovered, quite unexpectedly, that the Petersburg Academy of the Sciences did not consider him a member. He had received the diploma a long time before, but now his name was not on the rolls of membership. Apparently, they had never received his letter of acceptance of the nomination. At the time he had sent the letter with someone on his way to St. Petersburg (there was a draft of the letter among his papers), but a misunderstanding must have occurred. Pastor Collins, who lived in St. Petersburg, told of how he had received the request from Königsberg to deliver the letter to Princess Dashkov, but that the letter itself had never come into his hands.

Kant immediately wrote another letter of gratitude and sent it to J.A. Euler, the then secretary of the Academy. The letter reached the addressee, and another famous name was added to the rolls of Russian academicians.

The original of the letter is now in the archives of the Soviet Academy of Sciences.[229] Strangely enough, the letter cannot be found in any of the German editions of Kant's correspondence; according to the Academy edition, the letter has been lost. Therefore, we shall present the full text here:

> His excellency the chairman of the board and director.
> Most respected Sir!
> From the information I have been given by the secretary of the chamber to the duke of Hollstein, Mr. Nicolovius, on 6 July of this year, as he passed through Königsberg, I gather that my letter of gratitude to the President of the Russian Imperial Academy on the occasion of my election to the body on 28 July 1794 in Saint Petersburg never reached the addressee, and that as the result a substantial mistake in its registry has been made.
>
> As I was ignorant of the correct procedures I may have made a mistake by sending my letter of acceptance to the Academy (whose transmission to the authorities was assured me here by Mr. Collins through receipts for the postage) not through the Director but the then President, the Princess Dashkov, a mistake which can, I hope, be rectified by this apology and explanation.
>
> With the greatest respect I remain your honour's
>
> obedient servant,
>
> Immanuel Kant
>
> Königsberg
> 12 July 1797

A contemporary of Kant's, the mathematics professor and poet Abraham Kästner, tried to imagine what would happen if Kant's dream of perpetual peace ever became a reality. The result was the following epigram:

> They listened to the wise men
> And banned all war forever;
> All men now live in peace,
> But the philosophers — never!

It is true that the conflict of opinions rages nowhere else in science with so much fury and yet without the hope of reconciliation. Kant was familiar with Kästner's epigram and included it in his pamphlet "An Announcement of the Forthcoming Treaty of Perpetual Peace in Philosophy." Goethe thought this work stylistically "more Kantian than Kant himself." We know that the poet was extraordinarily sensitive to the philosopher's sense of irony.

But how can an end be brought to the strife in the camp of the lovers of wisdom, where schools of thought wage campaigns against each other like armies? In the histories of nations antagonistic interests erupt in annihilating wars and thereby bring about agreements establishing universal peace. So, too, in the history of philosophy antagonisms among systems ought to create conditions conducive to the establishment of fundamental universal principles. Only critical philosophy, immune to both dogmatism and scepticism, can effect the solution to the problem, in that it draws clear lines separating wisdom from knowledge. Wisdom is the law of conduct.

As far as philosophy as a doctrine of knowledge is concerned, things stand the same as in the doctrine of wisdom: a guarantee of peace can only be given when the duty of truthfulness is strictly observed. Of course, not everything is true which man considers to be true, but everything he says must be true. Lies are of two kinds: a conscious untruth, and a conviction unsupported by evidence. In the first instance, a lie is knowingly presented as truth; in the second, a claim is made on insufficient grounds. You must not lie! — that is the imperative of philosophy. "When the commandment: thou shalt not lie (even with the highest of motives) becomes a genuine principle of philosophy as a doctrine of wisdom, it will not only establish perpetual peace but secure that peace for generations to come."[230] But here fate takes an ironic turn: Kant's thought opened an era of heated conflicts such as had never before been known in the history of philosophy.

According to Kant, every new system of thought passes through three stages. At first it escapes notice; then it is rejected, and, finally, it is subjected to "improvements," made to fit individual interests. Kant had long since ceased to pay attention to his enemies: he was far more disturbed by his friends and followers. Salomon Maimon was the first to try and improve transcendental philosophy. He came from a hamlet in Lithuania and at the age of nine achieved the highest degree of Talmudic learning; he was married at eleven, became a father at fourteen, was very enthusiastic about the doctrine of Maimonides (and changed his surname) and left for Berlin in search of a European education. Devoid of any means of support he turned to the Jewish community for help, but they declared him a heretic and threw him out into the street. He begged his way through Prussia, arrived at Poznan and there managed to eke out a living. New charges of heresy forced him to leave Poznan. He returned to Berlin: this time fate brought him and Mendelssohn together. Now he began to study philosophy with a passion, but, as before, without a system. After reading the *Critique of Pure Reason* Maimon became a Kantian; following the custom of Talmudic scholarship he wrote a lengthy commentary, *On Transcendental Philosophy*. Marcus Herz sent the manuscript to his teacher for comments.

Kant was up to his ears in work and wanted to return the manuscript unread, but a brief glance at the text convinced him that what he had before him was not insignificant. He read the beginning carefully, appreciated the author's abilities and understood his major point: Maimon was trying to develop variations on a theme from the *Critique*. Maimon questioned the necessity of "thing in itself;" this was the first critique of Kant from the position of the "right." (That is how V.I. Lenin characterised attacks on Kant from the positions of subjective idealism; materialists criticised Kant from the "left.") Kant wrote a lengthy letter to Herz with an analysis of Maimon's position (never before had he written such a detailed epistle!); he declined to publish an evaluation of the work, but he did suggest a compromise: Maimon should defuse his critique and direct all of his powers to the point on which they are in agreement—a reform of metaphysics.

Maimon printed the work as it was, without altering a line. He sent his objections to Kant. Naturally, he received no answer. Then Bacon's works came into Maimon's hands and he wrote a treatise

Bacon and Kant, in which he compared the two philosophical reformers, drew arbitrary analogies and insisted that the positions of his teacher were wrong. The treatise appeared in *The Berlin Monthly* (1790, No. 2), and the author sent the issue to Königsberg. No answer. This did not prevent Maimon from sending Kant a detailed synopsis of his work *On the Soul of the World*.

And again there was no answer. Nor did Kant answer Maimon's following letter, in which he again presented his views of Kant's theory of knowledge and begged the teacher to comment on them and on the philosophical dictionary which he, Maimon, had just published. The last epistolary outcry came in November of 1792: ". . . I beseech you in the name of your morality not to deny me an answer. . . . Even if your answer is short, a sign of your attention would suffice." Maimon writes that he had just understood the reason for Kant's silence; Kant must have disapproved of his conduct. Yet his comparison in the essay on Bacon was wholly impartial. Now he begged Kant to explain several points of the transcendental aesthetics to him: "You claim that the representations of time and space are forms of sensibility . . . I, on the other hand, hold that that is not universally true."[231] And so he argued throughout the letter.

Kant complained to Reinhold that he could not understand "what Maimon actually hoped to accomplish with his revision of critical philosophy."[232] He had to let others straighten Maimon out.

Friedrich August Hahnrieder had written no learned treatises. He simply wanted to prove the reality of the categorical imperative in his life. Without completing his course of studies he went to Russia, there to seek his fortune; he was one of Kant's favourite students, and he carried with him a letter of recommendation from the philosopher. The able and energetic young man quickly obtained an officer's commission and was assigned to General Suvorov as an aide de camp. He took part in the second Turkish campaign (1787–1792), fought at Fokshani, Rymnik, Ismailia. The categorical imperative did not permit him to observe the wholesale breach of laws by the Russian empire with indifference, and he wrote a letter to the czarina about administrative mismanagement. He was summoned to St. Petersburg; there he was found guilty of slander and sentenced to a long term of imprisonment. Hahnrieder, a pupil of Kant's, but also of Suvorov's, escaped. In 1796 he returned, penniless, to his native land. Kant took a lively interest in the fate of his pupil. Hahnrieder was offered the job of a surveyor, which would take him to west Prussia. He, however, had other plans; in his opinion, only rough physical labour, would afford him an opportunity of living according to the dictates of the categorical imperative. He decided to become a cabinetmaker. He left for Berlin with a new letter of recommendation from Kant in his pocket. (The letter was addressed to Kiesewetter and described Hahnrieder's plans as paradoxical, but not at all fantastic.) It was not easy to find a master cabinetmaker who would willingly teach his craft to an intellectual desirous of learning the pleasures of the simple life: the situation was too ambiguous. Finally a master was found, but he charged an outrageous price for his services. Kant's Berlin friends and admirers agreed to finance the apprenticeship, and they provided Hahnrieder with the means for a modest life. And so the erstwhile Prussian student and Russian officer became a simple labourer.

He was satisfied. The saw and the plane felt good in his hands. The work was not exhausting; Hahnrieder could not break his ties with the cultural environment. He attended Kiesewetter's lectures, visited the publisher Biester, read Kant and corresponded with him. "Morality is not a chimera; you have proved that, I am convinced of this and ready to conduct myself in accordance with my convictions. I am strong; I had courage enough to stand up against the dangers and unpleasantnesses of Russia; I followed the path of my duty; can I allow my courage to desert me now, when I have so much less to contend with?"[233] Hahnrieder found time to enlighten his co-workers and to write his memoirs. (When they were completed, Kiesewetter suggested that they not be published. It was not that they were badly written—Hahnrieder wielded the pen just as well as his cabinetmaker's tools—but that Kiesewetter was afraid of unpleasantnesses: Hahnrieder's memory was too good and too outspoken.)

A year passed. Suddenly, Hahnrieder understood that the categorical imperative was incompatible with salaried labour. He did not want to be a simple machine in the hands of the rich, so that he might satisfy their thirst for sensual pleasure. Only an independent producer can live in obedience to moral laws. So Hahnrieder came up with a new solution: he was going to become a landowner. He went to the highest offices of the government, had himself classified as a peasant and obtained land in west Prussia. Soon he had a family and was at last at peace. In his last letter to Kant (31 July 1800), there is a tone of self-satisfaction in addition to the customary Hahnriederian exaltation:

"I think that I now stand as high as any mortal can; I can think of nothing greater than to work the land in freedom from another's caprice; I feel this happiness deeply and would change it for nothing in the world. My life is like a novel."

Sigismund Beck would not have made the same claim. Outwardly, his life resembled the life of his teacher—student, *Privatdozent*, professor. As for the life of the mind, the resemblance could have appeared complete: "From the theses appended to your dissertation I see that you have grasped my concepts better than many others." Beck answered: "I have studied the *Critique of Pure Reason* with the greatest attention, and I am as convinced of its truth as I am of mathematical axioms. The *Critique of Practical Reason* has been my Bible ever since it came out."[234]

Kant asked Beck to put together a compendium of his major works for a possible publication by Hartknoch. The three volumes came out between 1793 and 1796: *An Explanatory Exposition of the Works of Professor Kant, Undertaken at his Request.* The publication was the cause of a lively correspondence between the teacher and the pupil. At first Beck was timid (The *Exposition* was his first work); he followed Kant's suggestions carefully and thanked him profusely. But after the publication of the first volume Beck became bolder and began to teach the teacher. Beck could never understand the idea of the understandings' activity in the formation of concepts: he kept on insisting that they were "original attributions." Kant patiently kept explaining his point of view, but finally he could no longer restrain himself and declared: "I notice that while I write this I cannot wholly understand myself."[235] Beck did not catch the irony. The third volume of the *Exposition* came out under the title *The Only Possible Standpoint from which the Critical Philosophy Can be Evaluated*; Kant was enlightened in the true meaning of his own philosophy from Beck's letters. He once addressed Kant as the man "whose time to join his forefathers draws near"; he thought the formula was sufficiently respectful.

But Fichte caused Kant the greatest worry. He was the most able of his pupils—and the most fanatic. He sought no advice, asked no questions, shared no thoughts, and asked for no approval. The priest of truth—he proclaimed it in the name of all and in the full consciousness of his calling to set men on the paths of righteousness by the power of his word. "It is dangerous to cross Fichte," said his contemporary Anselm Feuerbach. "He is an untamed beast who brooks no opposition and considers every enemy of his nonsense a personal enemy. I am convinced that he could play a Mohammad, if the times were Mohammad's, and spread his truth with sword and dungeon, if his lectern were a throne."[236] In France he could have become the leader of the Convention: but he had been born in Germany, and so his fate was to be a university professor.

He never had to worry about contact with his audience; brilliant successes were followed by dismal failures. Sometimes students serenaded him, at other times they broke his windows. Fichte went his own way. He published one book after the other. He drove Kant's thought about the activity of consciousness to the extreme, to absurdity. The first principle of all truth, he held, was activity itself, an absolute, universal "I." By that he did not mean individualism: "I" for Fichte meant "we," he subordinated personality entirely to the interests of the whole, the society, the state. Kant's idea of the universal order of law underwent a thoroughgoing change in Fichte: the latter would have the state govern and supervise economic relations and through them the conduct of the individual. At first he was so enthralled by the French revolution that he wanted to move to France and become a Frenchman. Ultimately he was disappointed by the revolution, issued a call to the nation to fight Napoleon and declared that the Germans were superior to all other nations.

Fichte regularly sent his books to Kant, but the latter did not read them. He found out about Fichte's "science of knowledge" (*Wissenschaftslehre*) from reviews published in journals. The impression was not favourable; Kant concluded that Fichte's speculations resembled a pursuit of mirages: you think you have caught one, but there is nothing in your hand. Finally Kant read Fichte himself, which made him even more certain of his opinion.

Just as the erstwhile Jacobin Bonaparte rooted out the followers of the revolutionary dictatorship in France, Fichte went after Kant's "thing in itself" hammer and tongs. When Bonaparte established his empire he retained the tricolour. Fichte did not deny Kant; on the contrary, he maintained that the system he had constructed was the same as Kant's — only better and more complete. The impression could arise that the pupil had completed his teacher's work. Reinhold, the convinced Kantian, went over to Fichte. Kant called both of them "hypercritical friends" and was beside himself with indignation, but he did not discuss the issue publicly. His silence caused doubt; it was understood as capitulation, as an admission that there was nothing he had to say.

In January 1799 *The Erlangen Literary Review* wanted an explanation. It called on Kant to declare his views on Fichte's teachings. Kant's answer was short, but unambiguous: "I consider Fichte's theory of knowledge totally unacceptable as a system. His pure theory of knowledge is neither more nor less than plain logic, which on its principles does not deal with the material of cognition but abstracts from its contents as pure logic; an attempt to pluck a real object from it represents an exercise in futility and has therefore never been attempted." Furthermore, Kant said, it was incomprehensible to him that he should be considered to be only the author of an introduction to philosophy and not of a philosophical system; he always thought that the completeness of his thought was the best proof of its truth value. Everything he had written should be taken literally, not figuratively. The declaration ends with the following strong words: "There is an Italian proverb which says: Oh, Lord, save me from my friends, I can defend myself against my enemies. There are kind, well-meaning friends who behave incorrectly in order to advance our cause, . . . but there are also so-called friends, given to lies and perfidy, who seek to destroy us although their protestations are to the contrary (*aliud lingua promptum, aliud pectore inclusum gerere*); thev and their intrigues must be treated with the utmost caution. But be that as it may, thanks to its indefatigable tendency to satisfy reason in its theoretical as well as its moral and practical aspects, critical philosophy must feel convinced that no change of opinion, no improvement or reconstruction of its edifice represents danger. The system of critique rests calmly on secure ground; eternally unshakeable, it will always be essential to the highest goals of mankind."[237]

Kant's "declaration" came out in Jena. At the same time an academic scandal had erupted there, ending in Fichte's being accused of atheism and forced to relinquish his chair.

Fichte's friends grumbled that Kant had maliciously cast oil on the fire and demanded that he be rebuked. Fichte had the grace to answer with restraint: in public, that is. In his letters he poured out his soul. He said Kant had "three-fourths of a mind," called his doctrine "total nonsense," his conduct "prostitution" (letter to Karl Leonhard Reinhold of 29 September 1799); finally he averred that the wise man of Königsberg "who had never been that conversant with his own philosophy, now neither knew it, nor understood it" (letter to Schelling of 20 September 1799).

(It was a stroke of luck for Kant that Fichte was not familiar with the former's ironic admission in the letter to Beck, that he did not understand himself "sufficiently well." Heine, the jester, did not know of it either; in his book on German philosophy he dedicated a whole chapter to its "incomprehensibility." Hegel is said to have complained on his death-bed that only one man had understood him, and then to have added angrily: "And even he did not understood me." That is difficult to believe. Kant at any rate appears to have counted on the proper understanding of his work; had he thought otherwise he would not have taken refuge in irony.)

Strictly speaking there was no need for Kant to place distance between himself and subjective idealism by means of a public declaration. In the *Critique of Pure Reason* a whole section is dedicated to the clarification of his relationship to Berkleyanism. Critical philosophy as a whole presupposed the existence of the objective world which is given independent of our consciousness; man meddles in the affairs of the world, but the world could get along quite nicely without him. According to Kant philosophy is essentially a corrective to sound common sense.

After he completed the edifice of critical philosophy in the *Metaphysics of Morals*, Kant felt the need to express his thought in a more pregnant form. Simultaneously he wanted to answer the question which had for some time been the principal philosophical question for him: what is man?

The *Anthropology* (1798) was the last work which Kant prepared himself for publication. It represents the summation of his contemplations about man and philosophy in general. If we compare the *Anthropology* and the *Critique of Pure Reason*, much in Kant's last work remains incomprehensible. But, if we place it next to the rest of Kant's philosophy, then its content becomes clear (formally, the *Anthropology* is one of Kant's most lucid works). We have come to the end of the road. And, at the same time it is a new beginning: it is wholly fitting that the study of Kant should begin with the reading of *Anthropology*. The reader must follow the direction of Kant's thought in reverse: the *Critique of Pure Reason* should come at the end.

(After the *Anthropology* I suggest that the beginner read the *Metaphysics of Morals*, in order to become familiar with ethics and the theory of right, the alpha and omega of Kant's thought; then the reader should tackle the *Critique of Judgment*, where the foundations of his philosophy are established and aesthetics is given its due. The *Prolegomena* and the *Critique of Practical Reason* can initially be omitted.)

A first glance at the *Anthropology* reveals that its structure corresponds to the general structure of critical philosophy. The major part of the work breaks down into three sections corresponding to the three faculties of the mind: cognition, the feeling of pleasure, and the deciderative faculty. These three faculties provide the content of Kant's three *Critiques*.

In the *Anthropology*, the fundamental propositions of critical philosophy are brought to bear directly on the world of man: his experience, his strivings, and his conduct.

Kant considers anthropology from a "pragmatic" point of view. That means that physiology, e.g., what nature has made of man, remains outside the range of interest: what interests Kant is what man makes of himself and what other men make of him. Today we call this cultural or social anthropology.

For Kant, man is the most important thing in the world. He reigns supreme over all the other creatures by virtue of his self-consciousness. It is this which makes him an individual, a person. Regardless of the changes to which he may be subject, he remains one and the same person. From the fact of self-consciousness egoism emerges as a natural characteristic of man. A logical egoist considers it unnecessary to measure his judgment against the understanding of others, because he does not need that testing stone of truth. (By this Kant means that we should never forgo the use of that instrument because it ensures the rightness of our judgment.) The aesthetic egoist is satisfied with his own taste and thereby deprives himself of the possibility of further perfecting himself; he applauds his own stand and seeks the criterion of the beautiful only in himself. The moralistic egoist, finally, is one who limits all ends to his own being; he considers only that to be necessary which benefits him and sees the highest motive for action in his own happiness, not in the notion of duty. The philosophy of the Enlightenment, whose fundamental principle was the specific individual, cultivated rational egoism as the ground of all ethical conduct. Kant repudiates egoism and in all its aspects, because it is incompatible with reason. To it he opposes pluralism, a way of thought whereby man considers himself and his conduct not as the whole world, but only as part of it, he being a cosmopolite. To know men is to know the world. Kant demands that egoism be restrained and that mental activity be brought under the fuller control of reason. In the face of this all, he is enamoured of imagination. But it is one thing to call forth one's own inner voices and keep them under control; it is something else again when they appear unbidden and control us. That is either madness or a sign of its impending outbreak.

Yet consciousness uncontrolled by reason continues to pique Kant's curiosity. Can man have representations and remain unaware of their existence? Even in his youth, Kant had thought about this, calling such representations "dark" representations. Now he treats them with detail and vigour. In the darkness of consciousness a process can occur as complicated as that involved in artistic creation. Imagine a musician, Kant writes, improvising on the organ and simultaneously carrying on a conversation with a man standing at his side; one false note and the harmony is destroyed. When, however, the musician while playing a melody gives free rein to his imagination and does not reflect on what he should do in the next instant, everything might turn out fine, but, some might still wish he could have written the notation down.

How strong are these "dark representations"? What is their place in the house of the man's mind? Kant is not inclined to underrate their significance. The understanding is sometimes incapable of freeing itself from their influence, even when it considers them absurd and tries to work against them. That is the case with sexual attraction, for instance. As for the sphere of our unconscious representations, it is greater than we can imagine. It is practically without limits: "on the great map of our mind there are only a few illuminated points: that in itself can excite in us a sense of wonder about our own being; if only a higher power were to say: let there be light!, immediately, without the slightest activity on our part (as for instance when we take up a writer with everything that is in his mind) half the world opens before his eyes."[238] Judgments regarding "dark representations" and their role in the creative process constitute an important supplement to criticism's epistemology and aesthetics.

Another such supplement to the three *Critiques* can be found in the part of the *Anthropology* entitled "An Apology for Sensuality." In the *Critique of Pure Reason*, sensuality was treated offhandedly, but not with scorn; now the same subject is treated with more detail. The difference in principle between sensuality and intellect is underlined again (Leibniz and Wolff had seen the difference between all representations as being merely a matter of degree); the thesis that the senses do not deceive us is restated, supplemented by the claims that they cannot lead us astray, nor rule over us.

The doctrine of appearances in the *Anthropology* falls within the realm of morality. Moral appearance is a rule of conduct. Under the mask of politeness, a blackguard of course can be hiding. But wise nature has made man susceptible to illusion, so that he may more readily strive towards goodness. The more civilised the man, the better actor he is; he gladly plays the role he has been assigned, identifying with it and living it. In the end virtue is triumphant, because appearance turns into disposition. Decency (the self-control that keeps passions in check) is a useful illusion which maintains distance between the sexes; it is necessary to prevent the exploitation of one sex by another for the purpose of selfish pleasure. Sensuality cannot be subdued by force: it must be outwitted. Moral appearance is absolutely essential to social life.

Kant divides the senses into two groups. The higher (touch, sight, hearing) are based on mechanical activity and lead to "cognition of things outside of us." The lower, "merely subjective" senses (smell and taste) are the result of chemical reactions. Sense impressions are not dead, unchanging impressions of an activity from without. Their intensity depends on a series of circumstances in which the perceiving subject finds himself. Kant enumerates the ways in which sense impressions can be enhanced. (One could say that artists of all times and all nations have made good use of them.) First and foremost is contrast. Every stimulus is strengthened when it is perceived against the background of its opposite; a well-tended piece of land in the midst of a desert gives the appearance of a corner in paradise. Novelty also enhances sense impressions: hence the passion for the rare, the unusual, the unobtainable. The change of impressions also enhances them: life in the city and life in the country, work and rest, sociable conversation and play, the occupations with history, poetry, philosophy, and mathematics, all give new powers to the mind. One must not start with strong experiences and quickly reach their apex. Pleasure delayed is pleasure enhanced — especially by the imagination.

Imagination is the architect of Kant's most important philosophical edifice. The philosopher asked the question of what man is; he could have answered that man is a creature who creates culture with the help of a wondrous faculty — imagination. In the *Anthropology* imagination is a correlative of the senses. Kant says more precisely that it can be productive and creative, but that it works with the material and forms of intuition and itself creates nothing new. Imagination also has its limits, beyond which lie mere phantoms. It rests on accumulated experience. Kant cites a story from Helvetius about a lady who looked through a telescope and saw the heads of two lovers on the moon. A priest looked through the same telescope and said: "Oh, no, madam. Those are two spires on a cathedral."

In the *Critique of Judgment* artistic creation was considered as a unique gift which could not be compared with any other form of activity. The possessor of such a gift was called a "genius." In the *Anthropology* the realm of genius is extended to include science. Kant draws a distinction between "discovery" and "invention." That which is discovered already existed, unbeknownst to all (Columbus discovered America). Invention is the creation of something that did not exist before (gun powder was invented.) "Talent for invention is called genius."[239] Kant counts Newton among the geniuses. This is a characteristic detail: scientific and artistic creation, that is to say the theoretical and the aesthetic principles, have been brought closer together. Both depend on imagination.

But being brought closer together does not mean a total identity. In the *Anthropology*, as before, the specific nature of the aesthetic principle was set out as the activity of mediation. For that was the major accomplishment of the *Critique of Judgment*: it was shown that the aesthetic could not be reduced to either the cognitive, or the moral, yet at the same time the unbreakable bond between the beautiful and the true and the good was affirmed. The aesthetic is something other than knowledge and morality; it is a unique "bridge" between them. Now, we know that an affirmation cannot exist without its negation, a thesis without its antithesis. In the *Anthropology* the emphasis is on the antithesis. Specifically, we find there the introduction of the concept of "aesthetic cognition." This concept was not present in the *Critique*; there the beautiful was treated as a "play of the cognitive faculties" which are merely a preparation for the act of cognition. The close proximity of these two spheres opens up the possibility of inserting yet a third, mediating sphere between them, one adjunct to both the aesthetic and the theoretical principle, where cognition acquires an aesthetic tint and the aesthetic is also cognition. The adoption of such a sphere does not annul the original constructions of Kantian aesthetics; on the contrary, it follows logically from its propositions.

It is significant that this development occurs precisely in the *Anthropology:* here man steps into the realm of scientific enquiry and literature becomes one of the most important sources of knowledge. ". . . and so the characters drawn by Richardson or Molière must acquire their substantiality from the observation of actual commissions and omissions of man: although they may be exaggerated in degree, in quality they must be in accord with human nature."[240]

Aesthetic cognition is a unique mediating sphere between sensibility and understanding. Logic accuses sensibility of superficiality and particularity in knowledge. The understanding is accused of the opposite — a certain sterility and abstractness. "The aesthetic act, whose first requirement is popularity, cuts a new path whereby it is possible to avoid both deficiencies."[241] The path does not lead away from science. Quite the contrary: "beauty is the flower, science the fruit"; the feeling for the beautiful Kant defines as "part sensual, part intellectual pleasure."[242]

One of the major categories of Kant's aesthetics is defined very precisely in the *Anthropology:* it is the "feeling of pleasure," the foundation of which is judgment. In man the brutish instinct for pleasure has been tempered by culture and morality. The enjoyment which is reserved for man is synonymous with cultivation. "One of the ways in which we find pleasure is simultaneous cultivation: namely, the enhancement of the capacity to obtain more satisfaction in this way; science and the arts afford such satisfaction. Another way is attrition: it makes us less and less capable of further satisfaction. Whichever way man chooses: as has been said above, it is one of the most important maxims that pleasure should be pursued in such a way as to allow it ever to increase"[243] And then Kant turns to a young man and says: ". . . come to love your work; eschew pleasure, not to give it up altogether but to always have the prospect of it!"[244]

The famous proposition, "work is the best way to take pleasure in life," is presented in the *Anthropology.* The more you have accomplished, the more you have lived. The only way to be happy with your fate is to fill it with activity. Complete satisfaction is unobtainable anyway, and in that too, there is a deeper meaning: "Nature has made suffering the stimulus of human activity, the cause of man's irresistible drive towards a better life. In the last moment of life, the satisfaction with our last accomplishment is relative (partly because we compare ourselves with the fortunes of others, partly because we have in mind our own previous accomplishments); it is never pure and complete. To be (absolutely) satisfied during our life would mean an inactive peace and lack of incentive, or being dulled to experience and to the activity appropriate to it. Such a state is as incompatible with the intellectual life of man as the stilled heart with the life of the body . . ."[245]

Kant wrote these lines when he was seventy-five years old. He thought of only one thing — the completion of his work. And so he also brought newly found ideas into the *Anthropology*. More than anything else the book is a summation of his thought.

Kant is an eighteenth century philosopher. The century was coming to a close. The system of critical philosophy had been created and completed. Kant was never given to the feeling of self-satisfaction, but he knew that the most important work had been done. Perhaps the fact that the search had been brought to its end accounts for the quick demise of his powers.

He stood at the pinnacle of fame. He was a member of three Academies (Berlin, St. Petersburg, and, as of April 1798 the Academy of Siena). His biography had been completed long before. In 1792 Pastor Borowski had already sent him *A Sketch of a Future Biography of the Prussian Scholar Immanuel Kant*. The philosopher read the text, corrected inaccuracies and made a few additions, but he did not want to have it published during his lifetime. Another future biographer — Jachmann — decided to make his work easier: he submitted to Kant a detailed questionnaire containing fifty-six questions. Several of them had to do with the intellectual development of Kant's phiosophy: "44. Major points of change in philosophical orientation and the causes of move towards criticism. 45. In what order should one study the philosophical systems of ancient and contemporary philosophers? 46. What were their influences on your philosophy?" Other questions were more indiscrete: "33. Was there a woman lucky enough to have attracted your love and attention? 34. Which women had a positive influence on your sociability?" There was one question which threw light on Kant's relationship with the official church: "47. Did you ever observe the rituals of the Christian church, and when did you cease doing so?" If Kant had answered these questions we would have an exact biography at our disposal.

Fame has its negative side, too. Kant was pestered by people writing to him, by creators of philosophical systems, by the guardians of the purity of learning, and by plain swindlers. Some asked for spiritual support, others challenged him to public disputes, yet others asked for money. In Mecklenburg a certain "magister Kant" appeared who claimed to be the philosopher's son and sponged off the sage's well-meaning admirers.

Theodor Kant was no relation of the philosopher's. He appealed for help as the bearer of the identical surname: his farm had burned down and he estimated that he had lost five thousand talers. Could the great man, whom Europe holds in esteem, put himself in his shoes? Hard on the heels of this letter from Poznan a letter came from Sweden, in the spring of 1797. Karl Friedrich Kant was certain that the brother of his father, Lars Kant, must have been the philosopher's father. Karl Friedrich was planning to become an inspector, and therefore needed money. Could his highly respected cousin lend him a sum of eight or ten thousand talers for a few years? In summer the newly found cousin reminded Kant again of his existence and his plea. Then it appears that the Swedish bishop Lindblom was enlisted to lend support. The enthusiastic Latin epistle contained an important piece of information: the philosopher's father had been a noncommissioned officer in the army of Karl XII.

Lindblom's letter had to be answered. Kant was courtly and ironic: "Your grace's endeavours to cast light on my ancestry and the kindness with which you have let me know the results of your research deserve my heartfelt gratitude; even though it can be of no use to me, or anyone else."[246] Kant wrote that he knew perfectly well who his father was. As for his grandfather, he hailed from Scotland. (Here, as we know, Kant was wrong: his father was a Prussian, his great-grandfather came from Kurland.) Swedish connexions appeared to him highly problematic. Kant did not fail to quote the letter of his "cousin" which contained the request for a tidy sum of money; then he listed all of his closest relations: one sister, six nephews and nieces left after the death of his second sister, one brother and his four children.

Those who were closest to him by blood were far from him in spirit, but Kant gave them financial help whenever they needed it (more than a thousand talers a year). He left all of his estate to them, an equal share going to each. His last will and testament made according to the rules of the day was deposited with the university senate in February 1798. He left one thousand five hundred guldens to Professor Gensichen — the executor of his last will. Lampe, his servant, was not forgotten either. He was given a pension of four hundred guilders; in the case of his death his widow would receive two hundred guilders, and any surviving children would receive a lump sum of one thousand guilders. His estate consisted of the house and forty-two thousand nine hundred thirty guldens.

The spiritual estate consisted of handwritten manuscripts, drafts, and lecture notes. After the *Anthropology* Kant was unable to complete anything significant for publication. He left his lecture notes to his students: he gave the notes for the physical geography and pedagogy to Rink, the notes for logic and metaphysics to Jäsche.

There was a great deal of interest in physical geography. In 1797 Vollmer had offered to publish the lectures and to pay the fantastic sum of forty talers per proofsheet for doing so. Kant was amazed by the high price, but declined. When Rink got the manuscript he offered it to Vollmer on the condition that Vollmer publish six of his own works at the same time. That was an affront. Vollmer would have nothing to do with it, and he began to publish Kant's physical geography from his students' notes. A scandal was brewing. On 24 June 1801 Kant published his last piece — a protest in *Jenaische Allgemeine Literaturzeitung*. Kant declared that only Rink had the right to publish his lectures on physical geography. Vollmer then stated that Kant was no longer responsible for his actions, and that the protest had been inspired by Rink. He continued the publication.

Jäsche edited Kant's logic. The trouble with this edition was that Jäsche could not see the great changes in Kant's views over the course of time. Kant had taught logic throughout his teaching career, and he had always used Meyer's textbook. Each time he taught it he entered new notes in the book, which reflect the development of his thought. Jäsche saw it as one complete entity. Thus there are statements from different periods of Kant's work placed side by side in the text. For instance, the famous passage where the whole field of philosophical activity culminates in the question: "What is man?" That was apparently written in the last years of Kant's teaching. And then from a completely different area: aesthetics is defined as the science of the senses, and the sphere of beauty is thereby reduced to the lower levels of cognition. Not a word about the specific nature of the aesthetic, not a word about its mediating function. It is as if the *Critique of Judgment* had not been written, and Kant had returned to Baumgarten's positions.

Perhaps it was so: perhaps Kant had returned to the theories of his youth. But Kant had stopped lecturing in 1796; then he had published the *Anthropology*, in which he discussed the essential points of the ideas worked out in the third *Critique*. There can be no question of a return to the "pre-critical" position. Jäsche was simply not up to the task, and Kant could no longer supervise the edition.

1799 was the year of upheaval. The theologian Abegg, who had visited Kant the year before, had left a detailed account of the visit. He had dined with Kant twice. As always there were other guests, and the host led a lively and witty conversation. The talk was about political events and everything else. The new king, Friedrich Wilhelm III, had been in Königsberg. Kant did not like the fact that the king had been driven in a closed coach: he ought to have shown himself to the people on horseback. The philosopher had not taken part in the festivities. The queen wanted to see Kant and sent a servant to fetch him, but he would not answer the summons. Napoleon had embarked on a sea expedition; Kant was firmly convinced that he was going to occupy Portugal. (Even when the papers reported that Napoleon was in Egypt, Kant thought that it was merely a ruse. He believed that the occupation of Portugal was Bonaparte's major purpose, as a further blow against England.) The city gave the queen amber jewelry. Kant talked about the rare specimens of amber and about how the "gold of the Baltic" was obtained.

On the second occasion the conversation was about the peculiar behaviour of the Russian emperor Paul I and then about the future of England. Kant wanted it to turn republican, if King George were to be and remain the prince elector of Hannover. The Scots were abler than the English. Kant recalled the English proverb: drag a Scot through Europe in a sack, and when he returns home he will have learned all the languages of the continent. Everyone believed that the French republic was going to grow and flower; they believed it because that was what they hoped. Pulverised coal prevents the onset of putrefaction (the conversation had taken a radical turn): Kant remembered that he had once ridden in a postal coach in which there was a piece of roast venison which stank to high heaven; Kant sprinkled it with coal and all was in order. Eggs buried in coal powder retained their freshness longer. Coal and honey constituted an excellent remedy for toothache;

that should be told to the court preacher Schultz, who had long been complaining of toothache . . .

That was the year before. Now there were growing signs of senility. His walks were shorter, his step was unsteady, sometimes he fell to the ground. He went to bed earlier than before. Sometimes he fell asleep in his chair during the day. Company tired him out; he never invited more than two to dinner and it was becoming difficult to draw the host into a conversation. "Gentlemen, I am old and weak. You must treat me like a child."[247]

From among his friends Kant began to invite Deacon Wasianski, who had once studied under him. Wasianski was unprepossessing, reliable, caring. A jack of all trades, he could set aright whatever domestic mishap may have occurred — he could calm an irate Kant, fix the door, mend the clock, etc. (If Lampe took it upon himself to do such tasks, one could be certain of having to throw the mended piece out; the old soldier understood only what he had learned in the army.) Wasianski lived nearby and could drop in several times a day. Kant entrusted him with his household and financial affairs.

He got up early, as before. At five o'clock the philosopher was already at his desk. He wrote. He said that he was working on the major work of his life. The book was entitled, *The Transition from the Metaphysical Princples of Natural Sciences to Physics*. The manuscript grew. There were hundreds of pages. What are they about?

These are notes without any apparent direction. Some of the segments are quite lengthy, and in them a thought is consistently pursued, but more often we find unconnected fragments. Here, for example, are a few variants of the introduction.

> If the philosophical natural science (philosophia naturalis) is a system (and not merely an aggregate of fragments), then reason itself will separate its metaphysical original principles from the physical ones but these parts are heterogeneous and their completion is not so much a matter of development (*progressus*) as of transition (*transitus*) of metaphysical original principles of natural sciences into physics and neither the one nor the other for its own sake but for one another — This is not a jump from one realm into another, because in that case no necessary connexion of benefit to the whole occurs; rather it is a stand reason must take in order to bridge both shores with one step.[248]

Thus begins one variant.

> Natural science (*Philosophia naturalis*) is the science of the kinetic forces of matter in space. In so far as such a system rests on concepts and theorems *a priori*, it is called the metaphysics of nature. In so far as it is based on the principles of experience, it is physics. (. . .) The transition from the former to physics in order that both shores may be bound together is the task of the natural philosopher, because physics is the goal to which metaphysics strives, and in relation to which every concept is only a preparatory activity.[249]

So begins the second variant.

> The major undertaking of philosophical architectonics consists in preventing the boundaries between sciences from overlapping one another; it must establish accurate demarcations of each science, because without that the task of their further development becomes impossible.[250]

So begins the third introduction; none of them is carried out. Yet the thought is clear and free of contradiction.

Kant writes of solid bodies and of liquids, of movement, friction, generation of heat and the force of life. What is an organic body? An old question, but there is something new in the answer. An organic body, Kant argues, cannot obtain its organization only from the active forces of matter. Consequently, man must admit of the existence of an immaterial force, a force which is either part of the sensibly apprehended world, or something fundamentally different from it. Kant repeats such arguments time and again. Finally a statement is made of which we should take note: "The machine is a solid body whose construction is comprehensible only with reference to a purpose and whose form is analogous to a certain planned movement. If this form is imagined not as an actual intention, but as only a theoretical one, such a body is a natural machine. Organic bodies are therefore natural machines."[251] Here Kant develops ideas which he had already laid out in his letter to Sömmering on the "organ of the soul": life is a specific type of complex natural bonds. This anti-vitalist notion permeates the atmosphere of the epoch. Young Schelling will soon take it up when he begins to study speculative natural philosophy. Alexander von Humboldt formulated it as the exact result of an experiment. In his old age Kant seeks to keep abreast of developments in sciences. There is no senility here, nor a retreat into the "pre-critical" positions.

But years pass, and thought becomes confused. It jumps from one thing to another. Now the issue is no longer physics, but metaphysics, not nature, but being as such. Physics may be studied only after its philosophical foundations have been grasped. Therefore, one should begin with the transition from the metaphysical foundations of natural science to transcendental philosophy. Then it becomes possible to accomplish the transition to a general theory of experience, from nature to freedom, and, ultimately, to man as the being which unites both principles in itself.

The last file of reflections (from 1800) deals with an entirely abstract issue. First there is the question of God. What is God? Is there a God? God is a person possessed of all rights, in relation to whom man has no rights. It is a substance of highest being, free of all sensible representations. The concept of God stands in opposition to the concept of the world. The activity of the subject brings the two together. The thinking subject creates the world as the object of possible experience. This no longer sounds like critical philosophy: it seems to be a reference to Fichte. Or: "The knowledge of oneself as a person who constitutes himself as a principle and is the originator of that self."[252] Suddenly Kant calls his system "the theory of knowledge." Yet there is also the statement that I, man, am an external sensible object, part of the world. And among the philosophical propositions we find: "Wednesday. Peas with pork. Thursday. Dried fruit with pudding. Also, Göttingen sausage from Nicolovius."[253] Then again: God, the world, I the possessor of the world. Is it possible to wrest the world away from God? Transcendental philosophy furnishes no proofs for the hypothesis of God's existence. Philosophy is the love of a rational creature for the highest purpose of human reason. And then again everyday matters: "Yesterday evening Lampe hung my dressing gown behind the stove in the dining room so that when I put it on after dinner it would be warm, not cold. The cook fought furiously with Lampe: he orders her about and behaves like the master of the house. She, however, wants to play the role. *Homo homini lupus*."[254] And in another place: "Teach Lampe that he should not drink from morning till night; this week he shall not receive his wages, nor an increase."[255]

Kant's relationship with Lampe was never very close or idyllic. Lampe's stupidity in particular irritated Kant: for thirty years Lampe had been bringing the paper from the post office and still did not know its name. Once Lampe appeared before his master in a yellow coat (instead of the customary white coat with a red collar). Kant became angry. Lampe explained that the new acquisition had been purchased secondhand: he was getting married. Now, to Kant all this was news: that the servant was getting married, that he was a widower, that he had been married before. But Lampe's intended, as we know, was not forgotten in Kant's will.

Habit is a second nature. Kant had become used to his servant of many years; he tolerated his peculiarities and for fear of any change could not bring himself to get rid of him. Wasianski understood that matters could not go on like this. Admonitions fell on deaf ears. Lampe was a weak old man, like his master, and he got about with difficulty. Wasianski looked for a replacement.

Finally the day came when the master had had enough. Wasianski could not find out what had happened between the two, but Lampe was discharged on the spot. (He was given a year's salary and a pension of forty talers for life. A corresponding change was made in the will.)

The new servant, Johann Kaufmann, was a middle-aged man, even-tempered, a quick learner with a good memory. He adapted to Kant's habits with speed and the two got along very well. Wasianski supervised the first day of Kaufmann's service: at four o'clock in the morning he was at the house in Prinzessinstrasse. The philosopher got up at five and was somewhat surprised at the sight of two new faces instead of Lampe's. Kaufmann brought in the tea and Wasianski sat down at the table as well, but Kant was obviously ill at ease. He did not touch his cup.

> I sat right across from him. Finally he gathered himself and asked me very politely to sit so that he could not see me; for more than half a century he had been having his tea by himself. I did what he had asked, Johann left the room and came in only when Kant called him. Now everything was in order. Kant was in the habit, as I have already said, of drinking his tea by himself and pursuing his thought undisturbed. And even though he no longer read or wrote, the force of long habit was still strong in him, he could not tolerate anyone's presence without a mighty upheaval. The same thing happened when I called on him one lovely summer morning.[256]

In November of 1801 Kant took his final leave from the university. He had been retired with full salary. Now he very rarely left his house and he received no visitors. Very few were given leave to call on him. Among them was a young Russian physician, "follower of Kant's and a poet," who when he saw the old man fell to his knees and kissed his hand. Kant was confused and undone by the visit. The Russian showed up again the next day at the philosopher's house and asked the servant whether he could have something Kant had written. Kaufmann picked up a piece of paper from the floor of the study (it was a variant of the introduction to anthropology) took it to the parlour and gave it (with Wasianski's permission) to the waiting Russian physician.

He was beside himself with joy, kissed the paper repeatedly, took off his coat and waistcoat, folded them, put a taler on top and gave this all to Kaufmann.

When Kant's infirmity became known, unbidden guests began to call. A handsomely dressed woman insisted that she be given an opportunity to speak with Kant in private. First however she had to see Wasianski, who recognised her as a well-known adventuress. The lady declared that her husband had at one time given a dozen silver spoons and some gold to Kant; if she could not have them back, she would be perfectly happy with monetary compensation. Wasianski sent for the police and allowed the extortionist leave only after she had promised never to set foot in the house again.

Another unknown woman managed to make her way unobserved into Kant's study. Kant jumped from the desk, ready to defend his possessions. She wanted to know what time it was. Kant pulled out his watch, held it in his fist, and told her. The unknown woman thanked him, left the room, but then returned and said that actually the neighbour had sent her, and gave his name; he had to set his clock accurately and if the professor were to lend her his watch for a few minutes she would bring it right back. Kant let out a yell, and the lady retreated in haste. If there had been a scuffle, Wasianski remarked *à propos* of the event, the philosopher would have come off second best; for the first time in his life he would have been bested by a woman.

The torrent of letters dried up. Kant's friends knew of his condition. Kiesewetter wrote long after everyone else had stopped. He wrote less about philosophical problems than about turnips from Teltow: Kant was very fond of them and Kiesewetter regularly supplied him with the vegetable. Turnips are the major theme of their last exchange of letters. When Kiesewetter sent the usual quantity of turnips in November of 1801, he no longer wrote to Kant but simply put a shipping label on. Kant's last letter is dated in August 1801 — it is a note of thanks to Villers (to whom Bonaparte had once given four hours in which to explain Kant's philosophy), who had sent him two volumes on transcendental philosophy. The request that he be allowed to retire is written in someone else's hand; Kant only signed it. He dictates: "My strength ebbs with each passing day, my muscles are growing weaker; although I have never been ill, and am not ill now, I have not left my house for the past two years." That was in April 1802, in a letter to the husband of his niece, who had written to his famous relation a letter of introduction.

Kant could not walk without a supporting hand. Someone had to be in attendance at all times. Now his sister Barbara lived in the house (his brother had died). Now and then Kant sat down at his desk. He traced broken sentences in a shaky hand: "Transcendental philosophy is the totality of rational principles *a priori* integrated in a system (in a schema which displays the forms of cognition so that the material of cognition sets forth fully the forms of the principles involved) . . . Insomnia . . . In accordance with the principles of subjectivity the pain in the stomach is just that, a subjective experience. To give oneself over to the sight of starry skies — are they appearance or actuality?"[257]

Kaufmann helped his master keep a diary of sorts. In a special notebook he sets down the date, the menu, the guests (as usual two guests are invited to dinner). Among the notes are also occasional remarks of Kant's: "On Friday (the day after tomorrow) the 22nd I shall be eighty years old; I should like to invite my good friends. I.K."[258] That was dated 20 April 1803.

On the given day the guests gathered for a festive meal. But Kant did not feel well at the table. The noise of the conversation disturbed him, even though everyone tried to speak as softly as possible. "He only became himself when he was undressing in his study; we were alone and we talked about the presents he wanted to give to his servants. Kant could not be happy until all around him were happy. He brought a present to his servant every time he went out for a walk. I wanted to leave him in peace and took my leave in the usual fashion. He was always opposed to grand and unusual gestures, to all salutations of such kind, and if they were full of pathos he found them useless and funny. To my great surprise, he thanked me on this occasion in a manner which was totally out of keeping, his expressions bespoke clearly how weak he had grown."[259]

In October Kant's condition worsened. For the first time in his life he spent several days in bed. Then there was an improvement, and guests were again invited to dinner, which now took place in complete silence. Kant quickly emptied his plate and retired to bed. Sometimes he slumbered. In the evening he became restless. He had nightmares. Kaufmann slept in his room.

The last entry in the diary is dated 15 December. Kant had not been able to read since the fall. Now he was almost deaf. He no longer recognised his sister; the only one he did still recognise was Kaufmann. On 3 February 1804 he stopped eating. He sat at the table with his guests and was not able to eat.

Plato has left a moving description of the death of his teacher Socrates: execution by means of a medical procedure. Poison had to be drunk and then Socrates had to walk until his legs grew heavy. Then he had to lie down; when the cold reached the heart, it was the end. Socrates, falsely accused of atheism and the corruption of the young, chose death of his free will and refused the escape arranged by his friends. He went to death fully conscious and self-contained. Was that to be a challenge to his fellow-citizens who had despised his teachings? Or perhaps a call to the future generations to heed the voice of their conscience? For two thousand five hundred years mankind has been trying to unravel the riddle of this death.

Kant's death was as clear as his life. His duty had been done, then came the winding down and finally the end. The details are simple. On Saturday, 11 February, Wasianski spent the entire day at the bedside of the dying man. "I asked him whether he knew me. He could not answer and only offered me his lips to kiss. I was deeply moved, he offered me again his bloodless lips. I suppose he was taking leave of me and thanking me for the many years of friendship and assistance. I do not know whether he had ever kissed any of his friends."[260] Wasianski did not leave the bed. Kant's sister and a nephew were also in the room.

The agony lasted the day and well into the night. At one o'clock in the morning Kant came to and drank a few sips of wine mixed with water and sugar. He said: "It is good." And then he again slipped into unconsciousness. He did not come to again. Towards the morning he turned pale and became stiff. His eyes were open, but the light had gone out of them. The pulse could only be felt on the left hip. Wasianski knelt with his hand on the only part of the body where life still glimmered. Breathing grew weaker. His lips quivered and his breathing stopped. His pulse could still be felt for a few seconds, ever weaker; then the beats slowed and stopped. It was eleven o'clock on 12 February 1804. Kant was dead.

Kant had already made the arrangements for his interment in 1799. He wanted it to be on the third day after his death and as simple as possible: he wanted to be buried in an ordinary cemetery in the presence only of his friends and those closest to him.

Things turned out differently. The whole city took leave from Kant. The deceased was laid out for sixteen days. It was bitterly cold, and in the unheated parlour the remains of the philosopher who had become a skeleton in life resisted putrefaction. He was carried to the grave by twenty-four students, followed by the whole officers corps of the garrison and thousands of Königsbergers. The senate of the university waited for the procession at the cathedral, where the last eulogy was said. There was no priest.

Kant was buried in the "professors' crypt" in the north wall of the cathedral. It was an old structure, which became more and more dilapidated with the years. In 1809 it was pulled down and an open arcade was built in its place — the "Stoa Kantiana." At its end was Kant's gravestone and a bust.

The "Stoa Kantiana" survived until the end of the nineteenth century. In the eighties a neo-gothic chapel was erected over the grave of the philosopher. It is no longer there. In 1924, on the two hundredth anniversary of Kant's birth, a new monument was erected. It is still there.

As if by some miracle, Kant's monument was the only structure to survive the Second World War. Now surrounded by water, the island is clear of ruins except for the cathedral, where the grave is. On 22 April 1974, the two-hundred-fiftieth anniversary of Kant's birth was celebrated: philosophers came from all over the country and brought flowers.

By Way of an Epilogue

Man dies, thought endures. The philosopher achieves immortality when his word finds home in the world. Immortality knows no rules, the world is its home.

By way of an epilogue we should like to offer the reader three short studies of Kant's connexions with Russian culture. One is dedicated to a forgotten episode from Kant's life, the other two deal with Kant's influence in the sphere of Russian intellectual history. Together, they represent a summation of sorts.

When I was working on Kant's correspondence I came across an unknown name — Beloselsky. I searched through lexica — in the three large Soviet encyclopaedias I found nothing. Then I found the following in the *Russian Biographical Dictionary* (Saint Petersburg, 1908): prince Aleksandr Mikhailovich Beloselsky-Belozersky (1752–1808) was a highly educated man, member of the Academy of Sciences in Petersburg, of the Academy of the Arts, of the Academy of Classical Studies in Kassel, the Academy of Literature in Nancy, and of the

Institute of Bologna. Beloselsky was known as a diplomat and poet. He was the ambassador to the courts at Dresden and Turin. He wrote largely in French. He wrote the Russian libretto for *Olinka, or First Love*. There was nothing in the sources about his philosophical achievements.

Yet, Kant's interest is drawn precisely to Beloselsky's work in philosophy. In a draft of a letter to Beloselsky, Kant says: "Your grace has considered the metaphysical determinants of human cognition, the kind of work in which I have been engaged for some years, namely the consideration of human reason from another point of view, the point of view of anthropology . . ."[261]

We know that Kant was not given to flattery, and that he was not a great writer of letters. We remember that he did not answer the summons issued by Queen Louise, that he did not answer the mathematician Lambert who sent him his work on the philosophical method with a request for comments. Preparing his answer to Beloselsky, Kant worked up a draft in which he carefully weighed his formulations. What could this have meant?

I perused the thirteenth volume of the Academy edition of Kant's works, containing commentary on Kant's correspondence. It appears that Beloselsky had sent Kant his treatise *Dianyologie ou tableau philosophique de l'entendement* (*dianoia* in Greek means "the art of thinking," or reason), written in French and published in Dresden.[262]

Now the task was to discover the treatise. I called the State Library, in Leningrad, and in a quarter of an hour I had the answer: they had the second edition of the book, published in London in 1791, its call number was such and such. I could take the book home. That, of course, was a mistake: the book was out of print and therefore rare. But before they could find out that they had made a mistake, I had a xerox of it and was working on the translation. Then I discovered the first, the Dresden edition of 1790, in the Museum of Books. And a German translation, published in 1791 in Freiberg. (The *Dianyology* was apparently also published in English and Italian.)

In order to facilitate the book's reading, Beloselsky added a schema. The text of the treatise is essentially a commentary on the schema. Imagine five concentric circles, each one representing a cognitive faculty. In the centre there is "inert indeterminacy," the more possibility of cognition. The lowest sphere of cognition is the animalistic "torpor," where instinct and senses hold sway. Then comes the sphere of "simplicity, or judgment," the realm of common sense and intuition. Next to it we find the "sphere of the understanding," whose characteristics are clarity, consistency, and system. Above it, the "sphere of penetration, of transcendence," the grasp of the object as a whole. This is the sphere of philosophy, containing a number of similarities to Kant. For instance, Beloselsky speaks of play as a school of creative powers. And finally, there is the "sphere of the spirit," the realm of creative imagination, of genius. Spirit is the apex of cognition, it reveals more than man expects to see, it is an enemy of stagnation and compartmentalisation, its only limit is life itself: imagination must operate only with actual material, not with phantoms, otherwise its wings (so prettily represented in the schema) will easily carry it into the "realm of fictions." Between the "spheres" of cognition Beloselsky places the "realms of error," as if to point out frontiers beyond which cognition turns into its opposite. The "spheres" are narrow while the "realms" are broad concentric circles: the author is apparently convinced that in the world of human error, irrationality and stupidity greatly exceed man's capacity to increase and enlarge knowledge.

In Beloselsky's schema Kant found all the cognitive faculties he had described in the *Critique of Pure Reason.* Was Beloselsky familiar with Kant's work? What was actually Kant's influence on his work? It is hard to tell. After my first reading of the *Dianyology* I automatically assumed the Russian diplomat to be the disciple of the philosopher from Königsberg, who sent the fruits of his thinking about the works of the master to the master himself for commentary. How wrong I was! The mistake was understandable; I had found in a German book (K. Stavenhagen, *Kant und Königsberg*)[263] the mention of Beloselsky as one of Kant's Russian students. In the published list of matriculated students of the University of Königsberg there was no mention of Beloselsky's name. Perhaps the Russian aristocrat was not a regular student? There is no reliable information about Beloselsky's sojourn in Kant's city. There is no evidence that he had read Kant's works. Beloselsky's philosophical views developed under the influence of the French Encyclopaedists — D'Alembert, Voltaire, Rousseau, Montesquieu, Condillac, names we find on the pages of *Dianyology.* (Kant is not mentioned a single time.)

The difference between Kant and Beloselsky is two-fold: the major difference lies in their fundamental approach to the problems they treat; but their style of thinking is also different. For Kant, the faculties of cognition pertain to cognition itself: in various degrees they belong to all human beings. For Beloselsky, on the other hand, each "sphere" is a divison, a class of sorts, in which nature places man: within each division man can develop his faculties to the full, but he cannot go beyond the limits set by the system. Beloselsky sees social inequalities and decries them, but he sees the inequality of capabilities as natural and immutable. Man can only correctly grasp his position in this or that sphere and work on the development of the faculties with which he has been born.

Beloselsky adds another schema to his treatise: a "dianyological" classification of a number of world-famous men, divided into four groups according to the four-fold schema of "spheres." He does not bother to give reasons for his classification. One can only guess, for instance, why Madame de Pompadour finds herself in the "sphere of judgment," together with the philosopher Epictetus and the painter Dürer. In the "sphere of the understanding" we find Louis XIV, Luther, Poussin, Epicurus and Hume. In the "sphere of penetration" we find Cromwell, Calvin, Locke and Pascal. In the "sphere of the spirit" we find Peter the Great, Friedrich II, Leonardo da Vinci, Raphael, Michelangelo, Shakespeare, Plato, Descartes and Rousseau. Beloselsky's treatise here sinks to the level of an anecdote — amusing but shallow.

The *Dianyology* is not an accidental aberration in Beloselsky's opus. I cannot claim to have studied him thoroughly (although he deserves to be studied seriously, there is no doubt about that), but I have learned a thing or two, and I have managed to get hold of several of his unpublished works. In the manuscript *A Dialogue about Life and Death* Beloselsky speaks about the benefits of play, compares the talents of a player with the reason of a mathematician and refers in this relation to his *Dianyology*. At the end of the dialogue there is a sentence which corresponds to the seriousness of the treatise: Eat your fill, but not to excess. Aleksandr Mikhailovich Beloselsky was a wise and judicious man! The following aphorism shows how convinced he was about the immutability of human nature: believe, if you wish, that the mountain has moved, but never believe that man has changed his nature! (This is to be found in Soviet Central State Archives for Literature and the Arts (CGALI).)

Now, about the further search. There are two works about Beloselsky. One is Russian, done by A.A. Vereschagin, *The Apollo from*

Moscow, Petrograd, 1916 (an edition of three hundred!); the other is French, by A. Mazon, *Two Russo-French Writers*, Paris 1964.[264] Vereschagin describes the album of the Apollo from Moscow, as Beloselsky was called by his contemporaries. He was the "beloved of the Muses," a connoisseur of music and painting, a poet of Russian and French verse, preserved in the album. There are also letters from Voltaire, Marmontels and other men of note and autographs of Catherine II and Paul I. Through the documents Vereschagin weaves the story of Beloselsky's life. He speaks with fulsome praise about the prince's diplomatic dispatches from Turin: Beloselsky's objective analysis of the revolutionary events in France was the cause of the empress's displeasure and Beloselsky's subsequent recall from his post. Vereschagin takes Beloselsky to task on account of the latter's frivolous *Olinka:* the performance of the piece caused a minor scandal, and its author almost came to a bad end. Of *Dianyology* Vereschagin speaks only briefly and in negative terms. He thinks it is a "philosophical abracadabra."

Mazon uses a far greater number of sources than Vereschagin. He drew on archival material and published one of Beloselsky's unknown works for the first time.

His portrait of the young prince's tutor is very interesting. He was a Frenchman, a lawyer and a writer, member of the Berlin Academy of Sciences and a Jacobin. Mazon recounts the content of *Dianyology*, but relying on Vereschagin does not judge it positively.

But, let us put our trust in Kant. After the above-mentioned, spirited general characterisation of the treatise in the draft of the letter, an analysis follows of Beloselsky's schema of cognition. At first Kant repeats verbatim the arguments of the *Dianyology*. Then come refinements and digressions. And then the text breaks off. What was Kant's conclusion? What stopped him short? The original of the letter is not in the academy edition of Kant's works. Perhaps the letter was never sent?

We should not be hasty in drawing conclusions. Let us take another look at Mazon's book. Its appendix contains interesting material about the preparation of the unpublished Russian edition of the *Dianyology*. A part of the translation has been preserved, the title page bearing the censor's imprimatur of 3 January 1795 and the translator's foreword. As a curiosity, Mazon publishes the Russian translation of a letter from Kant to Beloselsky which was to have been included in the foreword of the planned Russian edition. He believes that the letter is not genuine, that it is the work of a flatterer from the prince's circle. Mazon doubts the genuineness of the letter because he could not find the German original in the archives, and "it proved impossible to uncover any trace of it in Germany."[265] Furthermore, someone had crossed out the translation.

A historian must be sceptical but he must also follow one rule: doubt must be doubted until it has been proven seven times! Mazon was overly hasty with his negative appraisal of the *Dianyology*. Mazon was also careless in his search for the traces of Kant's letter in Germany (he would have found it had he looked in the academy edition of Kant's works; there he would have found the draft of the letter whose opening phrases correspond to the finished variant). And, finally I have found out that Mazon inaccurately quoted the translation of the text from the archives.

Mazon refers to the archive of the Literary Museum. I was in for a disappointment: there is, and has never been, anything like it there. In the Theatre Museum the literary estate of the Princess Zinaida Bolkonsky (A.M. Beloselsky's daughter) was preserved; the material I wanted to see is part of it. Today the whole estate is in the Central State Archives for Literature and the Arts (fund 172, list 1, repository unit 153). I opened the file and found on the list of names of users the following note: "For Mazon." That means that the French Slavist did not work on the text himself: he was given a copy, and a carelessly made one at that — full of omissions and mistakes which must have made the already difficult text even more difficult to understand. Mazon's text, for instance, contains an omission: in the original the place is a large smudge. I took a magnifying glass and read: le bon sens. The Frenchman could not have been stumped. The translation of Kant's letter is literally crossed out. One could do nothing else with it. Only the German draft enables one to grasp its meaning. I studied the draft, tried to recall other passages from Kant, and reworked the Russian translation. Finally I had a Russian reconstruction of the text in modern philosophical terminology. Here, then, is the retranslation of the text from German, which might give us the sense of Kant's original letter:[266]

Kant to Prince Aleksandr von Beloselsky

I am in receipt of the valuable present — the excellent dianyology — which Your Grace was pleased to send to me last summer.

I have given two copies of the work to men who are capable of assessing its merits. I have spent the time since then in attempts to find an adequate expression of my gratitude, but various matters have come forth and caused this delay. Furthermore, I wanted to say something about what I have learned from the work, and this I shall do touching only upon the principal issues.

In order to set forth your profound analysis of cognitive faculties, according to scholarly rules, and adopt it using your concepts, I shall delineate the two separate realms, or areas, of our innate representational faculties (our natural metaphysical inclinations). The realm of the understanding is, in general, the faculty of thought, the realm of perception is the mere faculty of the senses.

The first of these realms consists of three spheres. The first sphere is the sphere of the understanding, or the faculty to understand, to construct concepts, and to work up perceptions and sense data. The second is the sphere of judgment, or of the faculty to apply these concepts to specific instances (*in concreto*), i.e., to bring into harmony with the rules of thought that which is actually constituted by the common sense (le bon sens). The third sphere is the sphere of reason, or the faculty of deducing particulars from universals, i.e., the faculty of judging according to principles.

When these three faculties of thought of the first realm are applied by analogy to the highest regulatory agency, i.e., reason, which serves to truly perfect man, and when they are integrated into a system, whose aim is the attainment of wisdom, we have before us the sphere of philosophy.

But, when they are integrated with the lower faculty (mere intuition), and especially with its most essential part, originality, which is constituted in the imagination (it is a power which does not obey laws slavishly but strives to create out of itself, as it is the case in the fine arts), then they constitute the sphere of the genius, the meaning of the word designating natural gift of the Creator.

In this manner I can discover five spheres. If, in the end, imagination annihilates itself by capricious activity, it turns into common foolishness or into nervous disorder; when imagination escapes the rule of reason and even tries to subjugate it, man leaves the estate (the sphere) of mankind and descends into the sphere of madness and phantoms.

I beg Your Grace to accept these undeveloped thoughts with kindness. I offer them as proof that I have thought about the content of your profound work,

I remain, etc.
Immanuel Kant

When we compare the final version of the letter with its draft, it becomes obvious that the philosopher had toned down his laudatory remarks. The penegyric to Beloselsky, for instance, is absent from the text. But Kant speaks to Beloselsky as an equal. And we must confess that without Kant's help it is difficult, perhaps impossible, to evaluate the merits of the *Dianyology* correctly. Examples of this are Vereschagin and Mazon.

But this is not the whole significance of Kant's letter. It contains a compact summary of Kant's philosophical position, given in his own hand. We are already familiar with its essential propositions. The understanding, judgement and reason are the three faculties of thought which encompass the world of philosophy when they are developed as a system. The understanding is the foundation of science, it develops concepts. Judgment applies these to the concrete instances of life. At first Kant had granted a fairly narrow realm to judgment, but in the end it became the mediating member of the system, that which makes art and culture possible. Reason is the controlling instance which regulates the work of the understanding and shields it from error; that is the sphere of morality, of the practical actualisation of philosophical principles.

The tripartite structure of philosophy is not a discovery of Kant's. The triad of the true, the good, and the beautiful goes back to Socrates. Kant subjected these concepts to an exhaustive scrutiny: separately and in relation to one another. The three critiques of Kant's correspond to Marx's idea of the three-partite appropriation of the world — the theoretical, the practical, and the practical-spiritual. Neither of them supercedes the other. They are equal, as they are equal in Kant's system. Primacy belongs to praxis.

Between Kant and Marx stand Fichte, Schelling and Hegel, masters of dialectic whose foundations were laid by Kant. But there is also a thread leading from the father of classical German philosophy directly to us. For instance, the idea of the synthesis of sensuality and the understanding in productive imagination (in epistemology), the emphasis on the role of duty in ethics, the analysis of the beautiful in aesthetics, and the project of perpetual peace in the theory of international relations.

Creation is the animation of Kant's philosophy. In it all of the intellectual capacities of man are brought together. However, we must not lose sight of how disturbed Kant was by the idea that the creative imagination could stray beyond the frontiers of the real and the rational. Man's mind, sound common sense, must always be on guard. After Kant there were attempts made to discredit "sound common sense." For Kant common sense is the guarantor of security, a compass as valuable as faith. Philosophy as a whole is actually nothing but a peculiar scientific touch-stone for common sense; common sense itself shields philosophy from scientific error.

Now we must return to the supposed conflict between Dostoevsky and Kant. Golosovker claims that "Dostoevsky was against Kant." He is wrong; he has apparently confused Kant with someone else. Perhaps with Hegel, whom Dostoevsky always judged with the utmost severity. "Hegel, this German pigmy, wanted philosophy to be the universal peacemaker"[267] — Dostoevsky was convinced that truth can be found in no abstract system of knowledge. To Hegel's proposition that "everything that is actual is rational, and everything that is rational is actual," the concept of reason unfolding itself in history, Dostoevsky opposed a no less decisive thesis: "Man can say everything about the history of the world, everything that the most extravagant imagination can think up: but he cannot say that it is rational." Hegel granted the spirit of the world the right to trample on the innocent flower. (The world spirit actually resides only in great men.) In *Crime and Punishment* Dostoevsky shows the inevitable result of the Hegelian proposition.

With Kant matters stood differently. Naturally, if Dostoevsky had been familiar with the "pre-critical" reasoning about fatal optimism, Kant would have been given a short shrift. Let us recall Ivan Karamazov's rebellion against God, as he tells the story of an eight-year-old boy who is torn apart by dogs: "A jester will perhaps say that it made no difference, that the child would have grown up to be a sinner, but this eight-year-old would never grow up, he was torn to pieces in the eighth year of his life."[268] The young Kant was once such a "jester."

But Dostoevsky probably knew none of that. At best, he read the *Critique of Pure Reason.* The ideas of the mature Kant he received through an intermediary, Schiller, as we have already mentioned.

N. Vilmont has written about this relation in his *Dostoevsky and Schiller.*[269] The book contains a few remarks about Kant, and it offers a judicious refutation of Golosovker's attack. If one sets aside the characterization of Kantian philosophy as "petty-bourgeois-eclectic" and the author's opinion that Dostoevsky was profoundly antihumanistic in his orientation, Vilmont's work is acceptable on the whole, because he convincingly integrates the Russian poet in the world of German humanism. But the final conclusion is stupefying: the resolute negation (in my opinion totally unwarranted) of the fact that "Kant and Dostoevsky were like-minded."

Where do I see the identity of their views? In the principal issue — the concept of the free personality. Kant's views we already know: freedom is the obedience to duty, and duty is the attendance to the happiness of others. Now let us hear Dostoevsky: "Does our salvation consist in the giving up of the personality? On the contrary, quite on the contrary, I say; not only must one not become an unperson, one must become a person in a sense much higher than is today held in the West. Understand me correctly: a free, supremely conscious and totally uncoerced sacrifice of the self to the good of others is in my opinion the sign of the highest development of the personality, of its highest power, its highest self-possession, the most sublime freedom of its will. Only the most completely developed personality is capable of giving up its life for others, of ascending to the cross or to the stake. A fully developed personality, profoundly aware of its right to be a personality, no longer fears for its own safety; it can do nothing else, that is to say, it cannot put itself to a higher use than the giving up of the whole self so that others may become equally self-determined and happy personalities."[270]

N. Berdyaev remarked that Dostoevsky was possessed of a "passionate feeling for personality."[271] The poet had not only studied paragraphs of Kantian critiques through mediation; Kant's fundamental propositions came to him directly: both Kant and Dostoevsky drank from the same source — the New Testament. Their understanding of Christian ethics was identical.

Christianity is not made of the same cloth. Dostoevsky, like Kant before him, rejected the Catholic variant as a false religion. Kant stood in opposition to the Protestant orthodoxy; the official Russian orthodoxy looked upon Dostoevsky with distrust. Both were, consequently, united in their unorthodoxy.

The teachings of Christ were for Kant and Dostoevsky the highest embodiment of the ideal of moral personality. Kant makes that statement in his ethics. Dostoevsky's philosophical masterpiece is *The Legend of the Grand Inquisitor*[272]; it deals with Kant's problem and enhances it in its most important moments.

The *Legend* unfolds in the Middle Ages. If Christ were to return to the earth, say the opponents of the Vatican, the Catholics would seize him and nail him to the cross. Ivan Karamazov phantasises upon the theme: the sixteenth century. Seville. The day before, one hundred heretics had been burned at the stake, in the presence of the king, his court and great masses of people. He appears quietly, without exciting notice, yet the crowd knows him at once.

The Grand Inquisitor, the cardinal executioner and theoretician of execution, knows him too. He orders that Christ be seized; then, when darkness settles over the city, he delivers what is simultaneously a speech, an accusation and a confession. He charges the arrival with making men unhappy by declaring them to be free, because man finds freedom unbearable.

The Inquisitor reminds Christ of the Satan's temptation in the desert: change stone into bread. Had he done that, mankind would have followed him like a grateful and obedient herd. "But Thou would not deny man his freedom and Thou rejected the tempter: what kind of freedom, you thought, can be bought by the price of bread? . . . Dost Thou know that centuries will pass and that mankind will proclaim through its wisdom and science that there is no crime and, therefore, no sin, only hunger? 'Feed us, and then demand that we be good!' that is what they will write on the banners which they will raise against Thee and under which they will pull down Thy temple . . . Thou promised them heavenly bread, but I ask Thee again: can this bread be equal to the earthly bread in the eyes of the weak, eternally sinful and eternally ungrateful human kind?"

Thousands, tens of thousands, seek heavenly bread. But what about the millions, the tens of millions, who are not strong enough to give up the bread of the earth for the bread of heaven? The executioner claims (and perhaps he also believes) that the weak are precious to him. They are sinful and rebellious, but in the end they will become obedient. "They will admire us and consider us gods because when placed in charge of them we are prepared to bear their freedom — the freedom which they fear — and because we are prepared to rule over them — that is how unbearable freedom will become to them in the end. But we shall say that we are obedient unto Thee and that we rule only in Thy name. We shall deceive them again, we shall not permit Thee to come unto us again. And this deception shall be our burden, our cross, for we shall have to lie."

For Dostoevsky, as for Kant, lying is the most grievous sin. One must never lie, not even out of "love of men." First of all one must not lie to oneself. The lie is the mother of all transgressions. The lie is the mother of fear. If you wonder "What should I do?", the answer is: refrain from lying first of all. The Grand Inquisitor takes Christ to task for failing to satisfy man's eternal yearning for an object of worship: "So long as man remains free he strives for nothing so incessantly and so painfully as to find someone to worship. But man seeks to worship what is established beyond dispute, so that all men would agree at once to worship it. For these pitiful creatures are not only concerned with finding what one or the other can worship but with finding something that all of them can believe in and worship; what is essential is that all may be together in it. From the beginning of time this craving for community in worship has been the greatest ill of man, both of the individual and of the species. For the sake of common worship they have slain each other with the sword. They have set up gods and challenged one another, 'Put away your gods and come and worship ours, or we will kill you and your gods!' And so it will be to the end of the world, even when the gods disappear from the earth: they will prostrate themselves before idols just the same . . ."

Man has no greater anxiety, the Grand Inquisitor says to Christ, than finding someone quickly to whom he can hand over his freedom. "But only one who can appease their conscience can take over their freedom. With the bread Thou wast offered an indisputable power: give bread, and man will worship Thee, for nothing is more convincing than bread. But if someone else gains possession of their conscience — oh! then they will cast away Thy bread and follow after him who has ensnared his conscience . . . There are three powers, the only three on earth, which are able to conquer the conscience of these impotent rebels and to hold it captive forever for their happiness — these forces are miracle, mystery and authority."

Christ resisted all three temptations of the devil. He refused to perform the miracle of turning stone into bread; he did not wish to come into the possession of the secret and throw himself from the roof of the temple so that the angels might catch him and bear him away; he refused to accept the highest authority, the supreme power over the realms of the earth. Faith needs no proof, as Dostoevsky explains the lesson of the gospel. "No proof is of any help in faith," as the "devil," the guilty conscience, whispers to Ivan Karamazov. "Nothing can be proved," continues Father Zossima, "but man can convince himself. How? By what means? By the experience of active love. Love thy neighbour in deed and without respite. As you come to succeed in loving, so shall you become convinced of the existence of God and of the immortality of your soul."[273]

We know the argumentation from the *Critique of Pure Reason* and *Religion within the Limits of Reason Alone*. Kant cast aside all logical proofs of the existence of God and all the traditional tenets of faith: miracle, mystery and the grace that issues from the highest authority. At the end of his philosophical journey Kant came at last to the recognition of love as the morally constitutive principle. Dostoevsky got there in the beginning. Perhaps at the Semenovsky square, waiting for execution, when "there was no more than a minute of life left"? In the execution of his moral revolution Kant came to honour man; Dostoevsky came to love him. Not all men together, the whole of humanity, but the one individual who stands next to you. And that is strangely enough the most difficult task.

"I love mankind, but I amaze myself: the more I love mankind in general, the less I love men in particular, that is to say, taken as individuals. In my thoughts I often had very particular notions about serving mankind, and perhaps I was really capable of going to the cross for man if it had been necessary; yet, I know from experience that I cannot spend two days in the same room with a man. I am barely together with someone else before he stifles my personality, my self-love, and limits my freedom. In twenty-four hours I can come to hate the best of all men: I hate one because he eats slowly, another because he has a cold and blows his nose incessantly. It seems that I become a misanthrope as soon as I come into contact with man. But it was also the case that the more I hated individual men the more devoutly I loved mankind in general." That, in Dostoevsky's view, is the confession of a perverted humanistic consciousness.

The Grand Inquisitor loves men in his own way. He, too, knows their weaknesses. He tells Christ that man needs miracles more than he needs God. "And as man cannot bear to be without the miraculous, he will create new miracles for himself, and will worship deeds of sorcery and witchcraft, though he might be a hundred times over a rebel, a heretic and infidel . . . We have corrected Thy work and have founded it upon miracle, mystery and authority. And men rejoiced that they were again lead like sheep, and that the terrible gift that had brought them such suffering was, at last, lifted from their hearts . . . They will be convinced that we are right, for they will remember the horrors of slavery and the confusion to which Thy freedom brought them."

Before the eyes of the Grand Inquisitor an alluring "new order" beckons, one in which throngs of people, bereft of freedom, gladly break their backs for the benefit of the ruling few: "Yes, we shall set them to work, but in their leisure hours we shall make their life like a child's game, with children's songs and innocent dances. We shall even allow them sin — they are weak and helpless — and they will love us like children because we allow them to sin. We shall tell them that every sin will be expiated, if only it is done with our permission, that we allow them to sin because we love them, and that we will take the punishment for these sins upon ourselves. And we shall take it upon ourselves, and they will adore us as their saviours who have taken on themselves their sins before God. And they will have no secrets from us. We shall allow or forbid them to live with their wives and mistresses, to have or not to have children — according to whether they have been obedient — and they will submit to us gladly and cheerfully. They will bring everything to us — everything, all the most painful secrets of their conscience — and we shall have an answer for everything, and they will believe our answer, for it will save them from the great anxiety and terrible agony they now endure making their free and personal decisions."

Dostoevsky opens social and psychological depths here which were unknown to Kant. This is not so astonishing: exactly a hundred years separate the *Critique of Pure Reason* and *The Brothers Karamazov*, and Dostoevsky's gaze is fixed at least fifty years into the future. He saw Catholicism, but he also foresaw the social mythology of the twentieth century which freed man of the chimera of conscience by transferring all responsibility onto the Führer, whereby no errors were admitted within the borders of the universal cult of ascetic self-sacrifice. Kant was certain that Christianity would put an end to the authoritarianism of the Old Testament. Dostoevsky sees that the idea of the free personality is threatened.

To Catholicism and the secularised homilies of the Old Testament Dostoevsky compares the Christian faith of the people: not the official cult, but the love of man made flesh. The Russian people have "suffered like Christ." The crucified Christ is the symbol of Russia; the symbol of the unfortunate, the humiliated, the annihilated. Dostoevsky's heart and mind belong to them.

In order to grasp the closeness between Kant and Dostoevsky one must make a careful distinction between the official church and Dostoevsky's faith. "Scoundrels have mocked my uneducated and backward belief in God. These fools have never dreamt that the denial of God could be so powerful as it is shown to be in the 'Inquisitor' and its preceding chapter; a denial of God to which the whole book provides a refutation. My belief in God is not the belief of a fool (a fanatic)."

For Dostoevsky religion is merely a "formula for morality." Contrary to Kant, he believes that "morality is rooted in religion." And to Kant's statement: "How terrible is a God without morality" Dostoevsky replies: "Conscience without God is terrible."

Kant's name appears frequently in the works of L.N. Tolstoy. The writer studied the philosopher in the original German and in translation: the older he grew, the more thoroughly he attended to the task. Sometimes he entered into polemics with Kant, more often he tried to support him. Towards the end of his life he came to the conclusion that they were in total agreement; but he continued to complain that Kant was difficult to read.

War and Peace was the cause of their first serious encounter. Tolstoy had decided that a clear philosophy of history was going to constitute the foundation of the novel. The *Critique of Pure Reason* could not be left out. In Tolstoy's consciousness, however, Kant had merged with Schopenhauer. "I read Kant," he remembered, "and understood nothing; I began to understand him only after I had read and re-read Schopenhauer, whom I had held in great esteem for a while."[274]

The categories of the philosophy of history are causality, freedom, and necessity. Tolstoy knew the Kantian antinomy of freedom and resolved it as Kant had done: man in himself is free, but as he is locked in a system of dependencies he loses his freedom in one way or another: "If we study man by himself," he says in the epilogue to *War and Peace*, "if we isolate him from his environment, all of his actions seem to be free. But if we see him in any relation whatever to what surrounds him, if we see him in connexion to all kinds of things — to another man talking to him, a book he reads, the work in which he is engaged, even the air he breathes or the light that falls on the objects about him — we see that each of these circumstances has its influence on him and thus directs his action in some way or another. And the more we perceive of these influences the less free he seems to us and the greater the impression we have of the necessity to which he is subject."[275]

Men pursue their contingent, personal, frequently petty and selfish purposes; the sum total of their pursuits, however, is something altogether different — it is something determinate, it is historical necessity. The historical necessity is the result of millions of more or less free wills, more or less contingent actions. This had already been recognized in the eighteenth century by Kant, among others; Tolstoy is telling us nothing new here. His merit as a thinker consists in his applying a known philosophical truth to Russian history, for instance to the events of 1812.

We read in *War and Peace*: "The vast majority of men at that period in history paid no attention to the general development of things, but let itself be guided by personal, contemporary and transient interests. And thus these men brought forth the greatest imaginable utility at that time to the totality."[276] The Moscow population as they were in flight were the last to think about the sense of their conduct; each fled for himself, each sought to preserve his life and property, or to relinquish them. And as a result, a "grandiose event" appeared as something inevitable, predetermined and subject to certain laws.

Tolstoy discoverd Kant's ethics in October 1887. How this happened we know from a letter to N.N. Strakhov: "I am very excited — these last few days I have been down with a cold, and as I could not write, I read the *Critique of Practical Reason* for the first time straight through. Tell me, please: have you read it? When? Were you impressed? Twenty-five years ago I put my trust in that talented dauber Schopenhauer (the other day I read his biography in Russian, as well as the critique of speculative reason, which is nothing but a polemic dispute with Hume and an introduction to his exposition of the fundamental theses of the

Critique of Practical Reason) and so came to believe that the old man had simply lied and that his central idea was only negation. For twenty years I had lived in that conviction and never thought of looking at the book itself. But such a relation to Kant is like taking the scaffolding around a building for the building itself. Is this only my error, or is it so general? I think that it is a general mistake. I looked through Weber's history of philosophy which happened to be at hand, and I saw that Mr. Weber does not approve of Kant's fundamental proposition that our freedom, which is determined by moral laws, is also a thing in itself (i.e., life itself), and that he only sees it as an excuse to run on and on about Fichte, Schelling and Hegel, and thinks that only the *Critique of Pure Reason* is worthy of note, i.e., he does not see the temple erected in the clearing but sees only the clearing, a fitting place for gymnastic exercises. Grot, doctor of philosophy, writes a paper about the freedom of the will, quotes some Ribaut and others whose definitions are exercises in nonsense and contradictions, and ignores Kant's definition; and we listen and blab and discover an already discovered America."[277]

A few days earlier Tolstoy had written to Biryukov:

> I am not quite well. I have a cough and am sitting at home and reading. The first reading of Kant's *Critique of Practical Reason* gave me a great deal of pleasure. How dreadful the fate of this remarkable work has been. It is the crown of his most profound intellectual activity, and no one knows it. If you do not read it in the original and I am still alive, I shall translate it and interpret it for you as best I can. If there is a biography of Kant in the public library, please take it out in my name and send it to me.[278]

From then on Tolstoy became a passionate populariser of Kant's ethics. For Tolstoy, Kant is a teacher of humanity together with Buddha, Confucius, Socrates, Christ, Mohammed, Rousseau and Skorovoda. He includes Kant's thoughts in the famous *Ring of Readings* and publishes them separately on two occasions. But he does not see the inner connexion between morality and cognition in Kant's thought. He sings praise to "the genius of Kant, who developed his ethics independently of metaphysics."

Working on his treatise *"What is Art?"* Tolstoy was drawn to Kant's aesthetics. Unfortunately it appears that his knowledge of the *Critique of Judgment* was limited to the reading of a German textbook on the history of aesthetics; he refers to it in the treatise. Consequently, Kant's views are incorrectly presented: beauty is only a unique form of disinterested pleasure[279]; the connexion of beauty to the true and the good, so important to the German philosopher, escapes Tolstoy's attention. He criticises this apparent position of Kant's, especially this "pleasure," which in the eyes of the Russian writer appears to be synonymous with the leisure activity of certain levels of society. Art for Tolstoy is a medium of human communication, accessible to all.

But, when one sets aside terminological niceties and pays heed to the heart of the matter, similarities between Kant's and Tolstoy's aesthetic concepts become apparent. In the manuscript of "On What We Call Art" Tolstoy compares art and play, and we know what is to follow: "Play is a necessary condition of the life of man, for the young for example, or for those who turn life into a celebration; it is the display of the surfeit of powers which are not expended in material activity. Art is a necessary condition of life for adults or old people, whose physical powers are used up in work or weakened by illness and old age. Two things are necessary to man, as a respite from the vicious circle of hard work, sleep and eating in which he gyrates from birth until death, like every living creature: these two kinds of entertainment — art and play — have been and will continue to be enjoyed by man . . . Art is a form of entertainment by means of which man experiences different feelings, and thus recuperates from the pain of living, without actively doing but by appropriating impressions."[280] What is the difference between Kant's "disinterested pleasure" and Tolstoy's "entertainment"? Each should judge according to his lights.

Every visitor to Yasnaya Polyana first goes to Tolstoy's grave. How different it is from Kant's grave! There gray gravestones against the background of ruins: life is gone, only the eternity of granite remains. Tolstoy's grave lies in the middle of greenery, in the forest, surrounded by grassy banks and flowers: it is as if the mortal remains were interred yesterday. There is no memorial, no cross, no tablet. Only the eternally inexhaustible life of nature reminds one of eternity . . .

In the house I asked to be shown the works of Kant which were in Tolstoy's possession. There is a twelve-volume edition of collected works, beautifully bound, new and undisturbed. These volumes were put on the shelves, however, the others had been thoroughly studied. The others contain the marks of sustained work: important pages are folded at the edges, sentences and whole sections are underlined, everywhere there are notes in the margins. Tolstoy read Kant with care, thought about him and lived his thought.

In the *Critique of Practical Reason* the well-known sentence about the starry skies above us and the moral law within us is thickly underlined. Tolstoy read the *Critique of Pure Reason* in a French translation: he paid special attention to the last sections. We remember Kant's insistence on the difference between opinion, knowledge, and belief. This passage is also underlined, and in the margin there is a comment: "Nonsense!" (Tolstoy was offended by the definition of belief as a simple subjective conviction.) But he agreed fully with Kant's notion that morality ultimately leads to God. "The good leads to the knowledge of God — always," so Tolstoy summed up Kant's remark; an analogous passage is given the highest mark — "5 +". Where Kant offers an apology for the common human understanding with the stipulation that "in the pursuit of the essential purpose of human nature the highest philosophy can bring us no further than the guidance that the most ordinary understanding provides" (B 859) Tolstoy has written: "Splendid."

In Tolstoy's waning years *Religion within the Limits of Reason Alone* became one of his favorite books. The Leipzig Reclamverlag pocket edition was always at hand. In the diaries we find frequent entries such as: "I read Kant and was moved," "very good," "I think so too," and so on. All pertain to Kant's philosophy of religion. Tolstoy was impressed by its moral fervour. Kant's understanding of God as the love of mankind was near and dear to him.

"Love is life. Everything but everything I understand, I understand only in so far as I love it. Everything is, everything exists only because I love. Love binds everything together. Love is God, and death means that I, a particle of love, return to eternal and universal love." Thus thinks Tolstoy's major figure, Prince Andrey Bolknosky, wounded in the battle of Borodino and ready to leave life. That is the answer of the Russian classic to the primeval questions of philosophy and art about the essence of man. The answer is permeated with Kant's spirit.

Man dies, thought remains. Thought assumes an independent life. It calls forth thought in others, becomes the possession of many. Thought penetrates the depths, opens up new realms of being and subjects them to the will and reason of those who live, dare, and must bear responsibility. The philosopher achieves immortality when he succeeds in articulating what has been heard, understood, and made real.

Chronology

1724	22 April. Immanuel Kant born in Königsberg.
1730	Enrollment in the elementary school.
1732	Enrollment in the Gymnasium.
1737	Death of Kant's mother.
1740	24 Sept. Kant enrolls in the University of Königsberg.
1746	Death of Kant's father.
	Thoughts On the Correct Evaluation of Life Forces goes to print. The work came out in 1749.
1747	Kant is tutor in the family of Pastor Andersch (Judtschen, near Gumbinnen).
1750	In the family of Major von Hülsen (Arnsdorf, near Osterode).
1753	In the family of Count von Keyserling (Rautenburg, district of Tilsit).
1754	Kant returns to Königsberg.
	June. *On Whether the Earth in its Rotation has Undergone Change.*
	August. *The Question of Whether the Earth has Aged from the Physical Point of View.*
1755	March. *A Universal Natural History and Theory of the Heavens.*
	17 April. Submission of the master's thesis *De igne.*
	13 May. Examination for the degree of *magister.*
	12 June. Promotion to the degree of *magister.*
	27 September. Defense of the thesis *Principiorum primorum cognitionis metaphysicae nova dilucidatio.*
1756	January–April. Two tracts and the book about the Lisbon earthquake.
	10 April. Kant's defense of *Monadologia physica.*

1757	Spring. *Syllabus of a Course in Physical Geography.*
1758	January (until July 1762) Kant a Russian subject.
	Spring. *New Science of Movement and Rest.*
	December. Kant fails to obtain Kypke's chair.
	14 December. An appeal to the empress Elisabeth.
1759	October. *On Some Observations about Optimism.*
1760	June. *Thoughts on the Premature Death of Johann Friedrich von Funk.*
1762	*The False Subtlety of Four Syllogistic Figures.*
	Herder attends Kant's lectures (until 1764).
	December. *The Only Possible Proof of God's Being.*
1763	*An Introduction of the Concept of Negative Quantities in Philosophy.*
1764	*On the Beautiful and the Sublime.*
	On the Illnesses of the Head.
	On the Clarity of Fundamental Propositions of Natural Theology and Morality.
1765	*An Announcement of Kant's Courses for the Winter Semester 1765/66.*
1766	*Dreams of a Spirit-Seer Made Clear by the Dreams of Metaphysics.*
	Kant becomes an assistant librarian in the royal castle.
1768	*On the First Source of Different Aspects of Objects in Space.*
1769	Invitation from Erlangen.
1770	Invitation from Jena.
	Kant's appointment to the chair of logic and metaphysics.
	21 August. Defense of the inaugural thesis *De mundi sensibilis atque intelligibilis forma et principiis.*
1771	Review of Moscati's work.
1772	21 February. Letter to Herz, containing plans for the *Critique of Pure Reason.*
	May. Kant gives up the post of the assistant librarian.
1775	*On Different Races of Man.*
1776/77	Two statements on *Philantropin.*
1778	Minister von Zedlitz urges Kant to accept the appointment at Halle.

1780 Kant becomes member of the university senate.

1781 May. Publication of the *Critique of Pure Reason.*

1783 *Prolegomena to Any Future Metaphysics.*

Review of Schulz's *Introduction to the Science of Morality.*

Kant buys a house.

1784 November. *Idea for a Universal History from a Cosmopolitan Point of View.*

December. *Answer to the Question: What is Enlightenment?*

1785 January and November. Review of Herder's *Ideas.*

March. *About the Volcanoes on the Moon.*

April. *Foundations of the Metaphysics of Morals.*

November. *On the Determination of the Concept of the Human Race.*

1786 *The Conjectural Beginning of Human History.*

Spring. *Metaphysical Origins of Natural History.*

Summer. Kant elected Rector Magnificus of the University.

October. *What Does It Mean, to Orient Oneself in Thinking?*

7 December. Kant elected to the Berlin Academy of Sciences.

1787 Second edition of the *Critique of Pure Reason.*

Letter to Reinhold, containing the tripartite schema of the philosophical system.

1788 January. *On the Application of Teleological Principles in Philosophy.*

Spring. The *Critique of Practical Reason.*

Summer. Kant again elected Rector of the University.

1789 Karamzin visits Kant.

1790 The *Critique of Judgment.*

Third edition of the *Critique of Pure Reason.*

1791 August. Fichte travels to Königsberg to meet Kant.

September. *On the Failure of Philosophical Enquiry into Theodicy.*

1792 April. *About Radical Evil.*

1793 Fall. *Religion Within the Limits of Reason Alone.*

September. *On the Saying: It May Be Correct in Theory, But It Does Not Work in Practice.*

1794	May. *On the Influence of the Moon on the Weather.*
	June. *The End of All Things.*
	28 July. Election to the Petersburg Academy of Sciences.
	12 October. Kant reprimanded by a royal order in council.
1795	*Perpetual Peace.*
1796	Response to Sömmerring's *On the Organ of the Soul.*
	23 June. Kant's last lecture.
1797	*Metaphysics of Morals.*
	July. *An Announcement of the Forthcoming Treaty of Perpetual Peace in Philosophy.*
	September. *On the Alleged Right to Lie Out of Love of Man.*
1798	4 April. Kant elected to the Siena Academy of Sciences.
	Fall. *The Strife Among the Faculties.*
	Anthropology From a Pragmatic Point of View.
1799	Public statement *contra* Fichte.
1800	Kant's last work prepared for publication by himself—an afterword to the Lithuanian-German dictionary.
	September. Jäsche publishes Kant's *Logic.*
1801	14 November. Kant asks to be relieved from his duties in the senate of the university.
1802	Rink publishes Kant's *Physical Geography.*
1803	Rink edits and publishes *On Pedagogy.*
	15 December. Last entry in the diary.
1804	12 February. Kant dies.
	28 February. Burial.
	May. Rink publishes Kant's competition thesis *On Developments in Metaphysics Since Leibniz and Wolff.*

Notes on the Sources

Original sources for quotations from Kant:

Kants gesammelte Schriften (Berlin: Preussische Academie der Wissenschaften, 1900).

Immanuel Kant: Briefwechsel (Hamburg: Meiner, 1972).

The following standard translations were consulted in the translation of specific excerpts from Kant's works:

Critique of Judgment, translated by James Creed Meredith. (Oxford: Oxford University Press, 1952).

Critique of Practical Reason, translated by Lewis White Beck. (Chicago: University of Chicago, 1949).

Critique of Pure Reason, translated by Norman Kemp Smith. (New York: Humanities Press, 1929).

Page numbers of sources refers to pages in the original edition of Kant's published works. In the case of the *Critique of Pure Reason*, page references are made to the second edition of the work.

Footnotes

FOREWORD

1. N. Gusev, *Dva goda s Tolstym* (Moscow, 1973), p. 190.

CHAPTER ONE

2. Johannes Scherr, *Deutsche Kultur- und Sittengeschichte* (Leipzig, n.d.), pp. 444, 447.
3. Franz Mehring, *Gesammelte Schriften*, "Die Lessinglegende" (Berlin, 1963), vol. IX, p. 113.
4. *ibid.*, p. 232.
5. Fritz Gause writes: "The earliest known ancestor of the philosopher is his great-great-grandfather, Richard Kant. He worked in an inn and married the daughter of the innkeeper from Heydekrug, Dorothea Lieder, or Liedert. A deed of the division of property among the inheritors and relations of his deceased wife from 1665 is extant; from the certificate of the interpreter in the proceedings it follows that Richard Kant knew German poorly. He had come from Kantweinen in the district of Prökuls, in northern Memel, which had been settled by the Kurlanders in the sixteenth century." Fritz Gause, *Kant und Königsberg: Ein Buch der Erinnerung an Kants 250. Geburtstag am 22. April 1974* (Leer/Ostfriesland, 1974), p. 11. Gause's source was an article by Hans and Gertrud Mortensen, "Kants väterliche Ahnen und ihre Umwelt," in *Jahrbuch der Albertus-Universität Königsberg III*, (n.p., 1953), p. 56.
6. *Immanuel Kant. Sein Leben in Darstellungen von Zeitgenossen.* Biographies by L.E. Borowski, R.B. Jachmann, and A.Ch. Wasianski. (Darmstadt, 1978), p. 163.
7. Karl Vorländer, *Kants Leben*, (Hamburg, 1974), p. 4.
8. *ibid.*, p. 26.
9. Kant, *Gedanken von der wahren Schätzung . . . , (Thoughts on the Correct Evaluation of Life Forces)* I, p. 24.
10. *ibid.*, I, p. 10.
11. *ibid.*
12. Kant, *Allgemeine Naturgeschichte und Theorie des Himmels . . . , (A Universal Natural History and Theory of the Heavens)*, I, p. 227.
13. *ibid.*, I, p. 230.
14. *ibid.*, I, p. 327, f.
15. *ibid.*, I, p. 306.
16. *ibid.*, I, p. 321.
17. *ibid.*, I, p. 366, f.

18. Letter to Johann Gotthelf Lindner of October 28, 1759 (17).
19. Kant, *Entwurf und Ankündigung eines Collegii der Physischen Geographie (Syllabus of a Course in Physical Geography)* IX, p. 427.
20. Kant, *Physische Geographie, (Physical Geography)* IX, p. 427.
21. Johann Wolfgang von Goethe, *Dichtung und Wahrheit* (Hamburg, n.d.), vol. IX, p. 30, f.
22. Kant, *Geschichte und Naturbeschreibung . . . des Erdbebens . . . , (History and physiography . . .)* I, p. 434.
23. *ibid.*, I, p. 460.
24. A. Bolotov: *Zhizn' i priklyucheniya Andreya Bolotova, opisannyie im samym dlya svoikh potomkov* (St. Petersburg, 1870), I, p. 623.
25. Maslovsky, *Russkaya armiya v semilyetnyuyu voynu*, (Moscow, 1888, 2nd ed.), appendix 14.
26. A. Stenbock-Fermor, *Der rote Graf*, (Berlin, 1953).
27. Bolotov, *op. cit.*, I, p. 631.
28. Bolotov, *op. cit.*, II, p. 63.
29. Kant, *Versuch einiger Betrachtungen über den Optimismus, (On some Observations about Optimism)* II, p. 34.
30. *ibid.*
31. Kant, *Gedanken bei dem frühzeitigen Ableben des Herrn Johann Funk, (Thoughts on the Premature Death of Johann Funk)* II, p. 42, f.

CHAPTER TWO

32. Kant, *Bemerkungen zu den Beobachtungen über des Gefühl des Schönen und Erhabenen, (Notes to the Observations on the Feeling of the Beautiful and the Sublime)* XX, p. 44.
33. Georg Friedrich Hamann, "Socratische Denkwürdigkeiten", in, *Sämtliche Werke*, (Vienna: Nadler, 1950), vol. II, pp. 57–82.
34. Goethe, Dichtung . . . , p. 514.
35. Kant, *Bemerkungen zu den Beobachtungen . . . , (Notes to the Observations)* XX, pp. 14, 15, 55.
36. Kant, *Preisschrift*, II, p. 290.
37. Immanuel Kant. From lectures 1762 to 1764. Based on Johann Gottfried Herder's notes, *Kantstudiuen 88*, (Cologne, 1964), p. 71, ff.
38. Johann Gottfried Herder, *Werke in 5 Bänden*, (Weimar, 1957), V, p. 267.
39. Kant, *Versuch über die Krankheiten des Kopfes, (On the Illnesses of the Head)* II, p. 266.
40. Kant, *Beobachtungen über des Gefühl des Schönen und Erhabenen, (Observations on the Feeling of the Beautiful and the Sublime)* II, p. 221.
41. Karl Vorländer, *Immanuel Kant. Der Man und das Werk*, (Hamburg, 1977, 2nd ed.), I, p. 194.
42. Kant, *Bemerkungen zu den Beobachtungen . . .* , XX, p. 73.

43. Kant, *Anthropologie, (Anthropology)* VII, p. 282.
44. *ibid.*, VII, p. 313, f.
45. *ibid., Anmerkungen*, VII, p. 311.
46. Kant, *Beobachtungen über des Gefühl des Schönen und Erhabenen, (Observations on the Feeling of the Beautiful and the Sublime)* II, p. 254.
47. Kant, *Bemerkungen zu den Beobachtungen* . . . , XX, p. 38, f.
48. Kant, *Reflexionen 2018/2021, (Reflections)* XVI, p. 197, f.
49. Kant, *Reflexionen 903*, XV.I, p. 395.
50. Kant, *Bemerkungen zu den Beobachtungen* . . . , XX, p. 45.
51. Friedrich's rescript of 16 November 1764 (34).
52. Bolotov, *op. cit.*, I, p. 705.
53. Letter from Johann Heinrich Lambert of November 13, 1765 (39).
54. Letter to Miss Charlotte von Knobloch of August 10, [1763].
55. *Immanuel Kant. Sein Leben in Darstellungen* . . . , pp. 96–103.
56. Kant, *Träume eines Geistersehers, (Dreams of a Spirit-Seer)* II, p. 356.
57. *ibid.*, II, p. 367.
58. *ibid.*, II, p. 317.
59. Reinhold Michael Lenz: "Als Sr. Hochedelwohlgebohrnen der Herr Professor Kant, den 21 sten August 1770, für die Professor-Würde disputierte", in, *Immanuel Kant zu Ehren,* Kopper and Malter, eds. (Frankfurt, 1974), p. 39, f. verses 3, 4, 11, 12.
60. Kant, *Reflexionen 680*, XV.I, p. 302.
61. Letter from Moses Mendelssohn of December 25, 1770 (88).
62. Letter to Marcus Herz (towards the end of 1773)(113 f.).
63. Letter to Marcus Herz of November 24, 1776 (150).
64. Letter to Marcus Herz, beginning of April 1778 (172).
65. Kant, *Vorlesungen über Pädagogik, (Lectures on Pedagogy)* IX, p. 443.
66. *ibid.*, IX, p. 497, f.
67. Vorländer, *Kants Leben*, p. 123.
68. *Immanuel Kant. Sein Leben in Darstellungen* . . . , p. 131.

CHAPTER THREE

69. Letter from Johann Caspar Lavater of February 8, 1774 (117).
70. Kant, *Reflexionen 1444*, XV.II, p. 630.
71. Vorländer, *Immanuel Kant*, p. 250.
72. Vorländer, *Kants Leben*, p. 89.
73. Letter from Abraham Freiherr von Zedlitz of March 28, 1778 (168, f.).
74. Kant, *KrV*, B XXXV.
75. Now and then Kant uses the term "thing in its own right." Gerhard Prauss says: "'Thing in itself' and 'thing in its own right' are merely shortened

forms of 'thing considered in its own right'." G. Prauss: *Kant und das Problem der Dinge an sich*, (Bonn, 1974), p. 20.

76. Kant, *KrV*, B, p. 334.
77. Kant, *KrV*, B, p. 83.
78. Kant, *KrV*, B, p. 190.
79. Kant, *KrV*, B, p. 85.
80. Kant, *KrV*, B, p. 294.
81. Andrey Bely: *Stikhotvorenia i poemi*, (Moscow-Leningrad, 1966), p. 307. (Poem "At the Window" from the cycle "Philosophical Melancholy")
82. Aleksandr Blok, *Sobranie sochineniy*, (Moscow-Leningrad, 1960), vol. I, p. 294. (Poem "Immanuel Kant")
83. Kant, *KrV*, B, p. 75.
84. Kant, *KrV*, B, p. 80.
85. Kant, *KU*, V, p. 197, note.
86. Kant, *KrV*, B, p. 108.
87. Kant, *Prolegomena & 36*, IV, p. 320.
88. Jacques Salomon Hadamard, *The Psychology of Invention in the Mathematical Field*, (n.p., 1945).
89. Kant, *Reflexionen 177*, XV.I, p. 65f.
90. Kant, *KrV*, B, p. 152.
91. Kant, *KrV*, B, p. 167.
92. Kant, *KrV*, B, p. 172.
93. Kant, *KrV*, B, p. 353.
94. Y.F. Golosovker, *Dostoevsky i Kant*, (Moscow, 1963), p. 36.
95. Kant, *Prolegomena & 52*, IV, p. 339.
96. Kant, *KrV*, B, p. 583.
97. Letter to Christian Garve of September 21, 1798 (780).
98. Blok, *Sobranie sochineniy*, (Moscow, 1971), vol. VII, p. 46.
99. F.M. Dostoevsky, *Polnoe sobranie sochineniy*, (Moscow, 1973), vol. VII, p. 161.
100. Golosovker, *Dostoevsky i Kant*, p. 87.
101. Heinrich Heine, "Zur Geschichte der Religion und Philosophie in Deutschland", in, *Sämtliche Schriften*, (München, 1971), vol. III, pp. 594–596.
102. Kant, *KrV*, B, p. 658.
103. Kant, *KrV*, B, p. 664.
104. Heine, *op. cit.*, vol. III, p. 604.
105. Kant, *KrV*, B, XXX.
106. Kant's play on words gets lost in the translation. *Cf*, I must therefore abolish knowledge in order to make room for belief, *Critique of Pure Reason*, n.tr., (London, 1930) p. XXX. Or: Je davais donc abolir la science pour faire place a la foi, *Critique de la raison pure par Kant*, n.tr. (Paris, 1864), vol. I, p. 22. The Russian translator uses either 'unichtozhit' (annihilate), or 'ogranichit' (limit). That is not a translation, but an interpretation.

Why did Kant play with words? Kant was openly ironic. In the same paragraph Kant says that one must avail oneself of the inestimable advantage of eliminating all objections by the use of "the Socratic method, namely by proving conclusively and clearly the untenability of the opposite proposition". (B XXXI) The "Socratic method" entails the use of irony. If one does not grasp the many-sidedness of Kant's proposition, a false understanding ensues. Korff recalls the famous sentence and interprets Kantian philosophy as a "rationalistic justification of irrationalism". (H.A. Korff, *Geist der Goethezeit*, (Leipzig, 1955), vol. II, p. 89)

But Vorländer had already referred to the "indeed ambiguous proposition". (Vorländer, *Kants Leben*, p. 159.)

107. Kant, *KrV*, B, p. 852.
108. Kant, *KrV*, B, p. 857.
109. Kant, *KrV*, B, p. 878.
110. Kant, *Prolegomena & 52 b*. IV, p. 340.
111. For reactions to the publication of the *Critique of Pure Reason* see, *Ein Jahrhundert deutscher Literaturkritik*, vol. III, *Der Aufstieg zur Klassik in der Kritik der Zeit*, n.ed., (Berlin, 1959), pp. 310–356; Herder to Hamann, p. 315.
112. Letter from Johann Heinrich Kant with postscript by his wife of September 10, 1782.
113. Letter to Christian Garve of August 7, 1783 (232).
114. *ibid.*, (228).
115. *Cf.*, *Prolegomena, IV*, p. 374, ff.
116. Kant, *Idee zu einer allgemeinen Geschichte in weltbürgerlicher Absicht*, (*The Idea of a Universal History from a Cosmopolitan Point of View*) VIII, p. 468.

CHAPTER FOUR

117. Kant, *Anthropologie, (Anthropology)* VII, p. 232.
118. Kant, *Idee zu einer allgemeinen Geschichte in weltbürgerlicher Absicht, (The Idea of a Universal History from a Cosmopolitan Point of View)* VIII, p. 24.
119. Review of Herder's *Ideen*, pt. I, VIII, p. 55. The whole polemic over Herder's *Ideen* is to be found in *Ein Jahrhundert*. . . , II, pp. 356–397.
120. *Herders Briefe*, (Weimar, 1959), p. 248.
121. *ibid.*, p. 245.
122. *Ein Jahrhundert*. . . , III, p. 367.
123. J.G. Herder, *Werke*, (Weimar, 1959), vol. 4, p. 268.
124. *ibid.*, p. 237.
125. Review of Herder's *Ideen*, pt. 2, VIII, p. 64.
126. *Cf.*, Vorländer, *Immanuel Kant*, I, p. 416.
127. Kant, *Grundlegung, (Foundations)* IV, pp. 393–394.
128. *ibid.*, IV, p. 404.
129. *ibid.*, IV, p. 404, f.
130. Kant, *KpV*, V, p. 30.

131. Kant, *Grundlegung*, IV, p. 398.
132. Kant, *Metaphysik der Sitten, (Metaphysics of Morals)* VI, p. 385.
133. Kant, *Grundlegung, (Foundations)* IV, p. 429.
134. Kant, *KpV*, V, p. 131.
135. Kant, *KpV*, V, p. 86, f.
136. Kant, *KpV*, V, p. 98.
137. Kant, *KpV*, V, p. 132.
138. Kant, *KpV*, V, p. 161.
139. Vorländer, *Immanuel Kant*, II, p. 307.
140. *ibid.*, p. 305.
141. Letter to Moses Mendelssohn of August 16, 1783 (233).
142. Kant, *Streit der Fakultäten, (The Strife Among the Faculties)* VII, p. 101.

CHAPTER FIVE

143. Kant, *Metaphysische Anfangsgründe . . . , (Metaphysical Foundations of Natural Science)* IV, p. 470.
144. Letter to Carl Daniel Reusch of May or June 1774 (133).
145. Letter to Carl Daniel Reusch of July 5, 1783 (218).
146. Letter to David Friedländer of November 6, 1787 (330).
147. Kant, *Reflexionen 2229*, XVI, p. 278, f.
148. Kant, *Über den Gebrauch teleologischer Prinzipien . . . , (On the Application of Teleological Principles)* VIII, p. 161.
149. Kant, *KU*, V, p. 306.
150. Kant, *KU*, V, p. 298.
151. This is how Horst Althaus understands the Kantian term: "Pleasure is a quality per se. The desire to obtain pleasure without considering the actual state of social mores, of purposes determined by the praxis of everyday living, or of the principles of art reception, always remains the first and aesthetically determining act. After Kant, Schopenhauer expressed this so decisively in his *The World as Will and Representation* that it should not be questioned as a fundamental maxim of aesthetics. Pleasure antedates the desire to exhibit a work of art; it invalidates all objections that reason may raise. In order that a work of art may be received, pleasure and the expectation of pleasure must override reason. Thus understood, pleasure is the one and only principle: it demands pleasure for its own sake and can therefore reject secondary appearances which are of ultimate significance to social groups, associations, even the whole species." Horst Althaus, *Ästhetik, Ökonomie und Gesellschaft*, (Bern and München, 1971), p. 309, note 2.
152. Kant, *Reflexionen 1521*, XV. 2, p. 885.

153. Kant, *KU, Einleitung*, V, p. 194.
154. Kant, *KU*, V, p. 217.
155. Kant, *KU*, V, p. 320, *Cf.* p. 314.
156. Kant, *KU*, V, p. 245.
157. Kant, *KU*, V, p. 261.
158. Kant, *KU*, V, p. 272.
159. Kant, *KU*, V, p. 274, f.
160. Kant, *KU*, V, p. 316, f.
161. Kant, *KU*, V, p. 304.
162. Kant, *KU*, V, p. 326, f.
163. Kant, *KU*, V, p. 326.
164. Kant, *KU*, V, p. 312.
165. Friedrich Schiller, *Versuch über den Zusammenhang der thierischen Natur des Menschen mit seiner Geistigen*, in, *Schillers Werke (Nationalausgabe)*, ed. by Benno von Wiese, (Weimar, 1962), vol. I, p. 40.
166. Letter to Friedrich Bouterwek of May 7, 1793 (637).
167. Schiller: *Über die ästhetische Erziehung des Menschen in einer Reihe von Briefen, op. cit.*, 2nd letter, p. 312.
168. *op. cit.*, 6th letter, p. 327, f.
169. *op. cit.*, 9th letter, p. 335.
170. *op. cit.*, 15th letter, p. 359.
171. *op. cit.*, 21st letter, p. 378.
172. *op. cit.*, 23rd letter, p. 384.
173. Kant, *Reflexionen 1485*, XV.2, p. 703.
174. Johann Wolfgang von Goethe, *Einwirkung der neueren Philosophie*, in, *Naturwissenschaftliche Schriften* (Hamburg, n.d.), vol. 13, pp. 27–28.
175. Vorländer, *Immanuel Kant*, II, p. 99.
176. A. Losev, V. Shestakov, *Istoriya esteticheskih kategoriy*, (Moscow, 1965), p. 326.
177. Goethe, *Anschauende Urteilskraft, op. cit.*, vol. 13, p. 30.
178. Kant, *Reflexionen 2978*, XVI, p. 597.
179. Kant, *KU*, V, p. 332.
180. Karl Marx, *Das Elend der Philosophie. Antwort auf Proudhons 'Philosophie des Elends'*, in, Karl Marx, Friedrich Engels, *Werke* (Berlin, 1964), vol. 4, p. 128, f.
181. Kant, *Muthmasslicher Anfang . . .*, VIII, p. 118, f.

CHAPTER SIX

182. N. Karamzin, *Briefe eines reisenden Russen* (Leipzig, 1799), pp. 57–63. (Further edition: Nikolai Mikhailovich Karamzin. *Briefe eines russischen Reisenden*, (Berlin, 1959); 2nd edition 1964.
183. Kant, *Religion innerhalb . . .*, VI, p. 168, note.
184. *ibid.*, VI, p. 97, f.

185. *ibid.*, VI, p. 128, note.
186. *ibid.*, VI, p. 125.
187. *ibid.*, VI, p. 126, f.
188. Imagination affects fear most powerfully. "The fear of fear", wrote A. Bely about A. Blok, "is the most real fear; he believed that Kant had fallen prey to such a fear — from then on Kant appears to him as a man terrified for all eternity". (A. Blok, *Sobranie sochineniy*, (Moscow, 1960), vol. I, p. 623) Blok's poem "I sit behind the screen" was originally entitled "The terrified one"; Blok used the antithetic coupling: "Shrivelled little Kant (Kantik) or before the wistful Kant (Kantische)", *op. cit.*, vol. 8, p. 70.
189. Kant, *Religion innerhalb . . . , (Religion within . . .)* VI, p. 170.
190. Kant, *Das Ende aller Dinge, (The End of All Things)* VIII, p. 330.
191. *ibid.*, VIII, p. 338.
192. *ibid.*, VIII, p. 338.
193. Georg Friedrich Wilhelm Hegel, *Werke*, (Frankfurt/M, 1969), vol. 17, p. 304 and *passim: Philosophie der Religion.*
194. Kant, *Muthmasslicher Anfang . . . , (The Conjectural Beginning . . .)* VIII, p. 112, f.
195. Kant, *Reflexionen 2487*, XVI, p. 390.
196. Kant, *Opus postumum*, XXI, p. 145.
197. Letter to Johann Caspar Lavater of April 28, 1775 (135).
198. Kant, *Metaphysik der Sitten, (Metaphysics of Morals)* VI, p. 437.
199. Vorländer, *Kants Leben*, p. 131.
200. Letter from Johann Gottfried Carl Kiesewetter of March 3, 1790 (440).
201. Letter from Kiesewetter of June 14, 1791 (511).
202. Letter to Ludwig Ernst Borowski, between 6 and 22 March 1790 (446).
203. Johann Gottlieb Fichte, *Von den Pflichten der Gelehrten. Jenaer Vorlesungen 1794/95*, (Hamburg: Meinner, 1971), p. 42.
204. Letter from Johann Gottlieb Fichte of August 18, 1791 (515).
205. Letter from Fichte [of September 2, 1791] (520).
206. Letter to Johann Erich Biester of May 18, 1794 (667).
207. Kant, *Das Ende aller Dinge, (The End of all Things)* VIII, p. 337.
208. *ibid., Anmerkung*, VIII, p. 331.
209. Letter from Joachim Heinrich Campe of June 27, 1794 (674).
210. *Cf., Proceedings of the Imperial Academy of Sciences*, (St. Petersburg, 1913), vol. IV, p. 386 (in Russian).
211. Order in council, Friedrich Wilhelm II, of October 1, 1794.
212. XII, p. 406.
213. Letter to Carl Friedrich Stäudlin of December 4, 1794 (687 f).
214. Kant, *Anthropologie*, VII, p. 30.
215. *ibid.*, VII, p. 35.
216. *ibid.*, VII, p. 42.
217. *ibid.*, VII, p. 41.
218. Kant, *Logik*, IX, p. 25.

CHAPTER SEVEN

219. Kant, *Zum ewigen Frieden, (Perpetual Peace)* VIII, p. 343.
220. *ibid.*, VIII, p. 369.
221. Quoted by Thomas Bruns, "Das politische Kantbild in Frankreich," in, *Akten des 4. Internationalen Kant-Kongresses, II, 2,* (Berlin, 1974), p. 649.
222. Vorländer, *Immanuel Kant,* II, p. 256.
223. Kant, *Metaphysik der Sitten, (Metaphysics of Morals)* VI, p. 321, f, and note.
224. Kant, *Streit der Fakultäten (The Strife among the Faculties)* VII, p. 85.
225. Kant, *Reflexionen 1520,* XV.2, p. 879.
226. Kant, *Über den Gemeinspruch: Das mag in der Theorie richtig sein, (On the Saying: It May be Correct in Theory but it Does Not Work in Practice)* VIII, p. 304.
227. Kant, *Streit der Fakultäten,* VII, p. 93, f.
228. Kant, *Metaphysik der Sitten,* VI, p. 350.
229. The letter is deposited under the call sign F.I., op.3, d.72 N 200. The Russian translation was published for the first time in *Studies in the History of Astronomy,* n.a. (Moscow, 1956), p. 373. The German original was published in, *Kant-Studien* (n.p., 1975) vol. I. One place in the letter must be clarified. Kant writes that he may have committed an error by sending the letter to the Academy not "through its director, but through its then president, Princess Dashkov." The president of the Academy at the time was (nominally) K.G. Razumovsky, who was abroad; Princess Dashkov was its director until November 1796. Her position was clearly stated in the diploma, but Kant did not notice it. To address I.A. Euler as the director, was another mistake; he was the recording secretary of the Academy.
230. Kant, *Verkündigung des nahen Abschlusses eines Tractats zum ewigen Frieden . . . , (An Announcement of the Forthcoming Treaty . . .)* VIII, p. 422.
231. Letter from Salomon Maimon of November 30, 1792 (606).
232. Letter to Carl Leonhard Reinhold of March 28, 1794 (662 f.)
233. Letter from Friedrich August Hahnrieder of December 3 1796 (532).
234. Letter from Jacob Sigismund Beck of October 6, 1791 (677).
235. Letter to Jacob Sigismund Beck of July 1, 1794 (677).
236. Kuno Fischer, *Fichtes Leben, Werke und Lehre,* vol. VI of *Geschichte der neueren Philosophie* (Heidelberg, 1900), p. 133.
237. Kant, *Erklärung in Beziehung auf Fichtes Wissenschaftslehre* (Declaration . . .) XII, p. 397.
238. Kant, *Anthropologie, (Anthropology)* VII, p. 135.
239. *ibid.,* VII, p. 224.
240. *ibid.,* VII, p. 121.
241. *ibid.,* VII, p. 146.
242. *ibid.,* VII, p. 239, 249.
243. *ibid.,* VII, p. 236, f.
244. *ibid.,* VII, p. 237.
245. *ibid.,* VII, p. 235.
246. Letter to Jacob Lindblom of October 13, 1797 (751).

247. *Immanuel Kant: Sein Leben in Darstellungen* . . . , p. 235.
248. Kant, *Opus postumum* XXI, p. 174, f.
249. *ibid.*, XXI, p. 176, f.
250. *ibid.*, XXI, p. 179.
251. *ibid.*, XXI, p. 211.
252. *ibid.*, XXII, p. 54.
253. *ibid.*, XXI, p. 71.
254. *ibid.*, XXI, p. 121.
255. *ibid.*, XXII, p. 119.
256. *Immanuel Kant: Sein Leben in Darstellungen* . . . , p. 263.
257. Kant, *Opus postumum*, XXI, p. 3, f.
258. *Kant-Studien 1970*, no. 1, p. 96.
259. *Immanuel Kant: Sein Leben in Darstellungen* . . . , p. 273.
260. *ibid.*, p. 301.

BY WAY OF AN EPILOGUE

261. Letter to Aleksandr prince von Beloselsky (draft), summer of 1792, (573).
262. *Dianyologie ou tableau philosophique de l'entendement*, par le prince Beloselsky (London, 1791). German edition: *Dianyologie oder philosophisches Gemälde des Verstandes*, von dem Fürsten Beloselsky. Translated from the French (Freiburg and Annaberg, 1791).
263. K. Stavenhagen, *Kant und Königsberg* (Göttingen, 1949) p. 38.
264. A. Mazon, *Deux Russes ecrivans francais* (Paris, 1964).
265. Mazon, *Deux* . . . , p. 80.
266. This is a translation from the first publication of the text in German. The original is deposited in the Central State Archives for Literature and the Arts (CGALI), under the call sign: Fond 172, op.I, ed.chr.153, list 21–25. I am indebted to Sigrun Bielfeldt (Tübingen) and Steffen Dietzsch (Berlin), who have helped me greatly with the retranslation.
267. *Neizdanny Dostoevsky* (Moscow: Literaturnoe nasledstvo, 1971), vol. 83, p. 404.
268. F.M. Dostoevsky, *The Brothers Karamazov*, trans. Constance Garnett (New York: Random, 1950), p. 300.
269. N. Vil'mont, *Velikie sputniki* (Moscow, 1966), pp. 141, 276.
270. F.M. Dostojevskij, *Winterliche Aufzeichnungen über sommerliche Eindrücke*, trans. Svetlana Geier (n.p.: Rowolt, 1962), Russ, Lit. vol. 8.
271. N. Berdyaev, *Mirosozertsanie Dostoevskogo* (Parizh, 1968), p. 52.
272. "The Grand Inquisitor" in F.M. Dostoevsky, *The Brothers Karamazov*, trans. Constance Garnett, (New York: Random, 1950), pp. 292–314.
273. L.N. Tolstoy, *Polnoe sobranie sochineniy* (n.p., n.d.), vol. 69, p. 24.

274. L.N. Tolstoy, *Polnoe* . . . , Epilogue, 2nd part.
275. L.N. Tolstoy, *War and Peace*, trans. Rosemary Edmonds, (Baltimore: Penguin, 1957).
276. L.N. Tolstoy, *Polnoe* . . . , vol. 64, p. 105, f.
277. *ibid.*, p. 102.
278. L.N. Tolstoy, *Polnoe* . . . , vol. 30, p. 57.
279. L.N. Tolstoy, *Polnoe* . . . , vol. 30, p. 253.

GPSR Compliance

The European Union's (EU) General Product Safety Regulation (GPSR) is a set of rules that requires consumer products to be safe and our obligations to ensure this.

If you have any concerns about our products, you can contact us on ProductSafety@springernature.com

In case Publisher is established outside the EU, the EU authorized representative is:

Springer Nature Customer Service Center GmbH
Europaplatz 3
69115 Heidelberg, Germany

Batch number: 10725993

Printed by Printforce, the Netherlands